TIME IN POWERS OF TEN

Natural Phenomena and Their Timescales

TIME IN POWERS OF TEN
Natural Phenomena and Their Timescales

Gerard 't Hooft · Stefan Vandoren

Utrecht University, The Netherlands

Translated by Saskia Eisberg-'t Hooft

 **World Scientific**

NEW JERSEY · LONDON · SINGAPORE · BEIJING · SHANGHAI · HONG KONG · TAIPEI · CHENNAI

Published by

World Scientific Publishing Co. Pte. Ltd.

5 Toh Tuck Link, Singapore 596224

USA office: 27 Warren Street, Suite 401-402, Hackensack, NJ 07601

UK office: 57 Shelton Street, Covent Garden, London WC2H 9HE

British Library Cataloguing-in-Publication Data
A catalogue record for this book is available from the British Library.

Translated by Saskia Eisberg-'t Hooft from the original Dutch edition *Tijd in Machten van tien — Natuurverschijnselen en hun tijdschalen.* © 2011 Veen Magazines, Diemen.

TIME IN POWERS OF TEN
Natural Phenomena and Their Timescales

Copyright © 2014 by Gerard 't Hooft and Stefan Vandoren

ISBN 978-981-4489-80-5
ISBN 978-981-4489-81-2 (pbk)

Printed in Singapore by Mainland Press Pte Ltd.

Contents

PART II

Foreword *by* Steven Weinberg

Ordinary human experience spans a range of times from seconds to decades, the longest intervals of time a mere billion or so times longer than the shortest. But the progress of science has been marked by the scientist's growing familiarity with time intervals that are very much longer, or very much shorter, than those that are experienced in our lives.

Around 150 BC the Greek astronomer Hipparchus observed that the position of the Sun at the time of the autumnal equinox was gradually changing, at a rate that would take the equinoctal Sun completely around the zodiac in about 27,000 years. Newton later explained this precession of the equinoxes as an effect of a slow wobble of the Earth's axis of rotation, caused by the gravitational attraction of the Sun and Moon for the equatorial bulge of the Earth. The Earth's axis is now known to make a complete turn in 25,727 years. Hipparchus had done the first serious scientific calculation of a time interval very much longer than a human lifetime, and found a result that was off by only about 5 percent.

In this century we have become used to much longer intervals of time. From the relative abundance of isotopes of uranium we can infer that the material of which the solar system is made was formed in an exploding star about 6.6 billion years ago. Looking farther back, by observing the way that galaxies now rush apart we can infer that 13.8 billion years ago the matter of the universe was so compressed that there were no galaxies or stars or even atoms, only a hot thick gas of elementary particles.

The extension of our experience to very short time intervals has been even more dramatic. By observing phenomena like diffraction that are associated with the wave nature of light, it became known early in the nineteenth century that a typical wavelength of visible light is about 0.3 ten-thousandths of a centimeter. Light was already known to travel at a speed of about 300,000 kilometers per second, so the period of the light wave, the time it takes light to travel one wavelength, was known to be about 10^{-15} seconds (a quadrillionth of a second). This is not very different from the time (to the extent that a classical description is relevant) that it takes electrons in atoms to make one complete circuit of their orbits.

Modern elementary particle physics deals with time intervals that are very much shorter. The lifetime of the W particle (the heavy charged particle responsible for the weak force that allows neutrons to turn into protons in radioactive nuclei) is only 3.16×10^{-25} seconds, not long enough for a W particle traveling near the speed of light to cross the diameter of an atomic nucleus.

What I find truly remarkable is not just that scientists have come to confront these very long and very short intervals of time. It seems to me even more amazing that our experiments and theories have become sufficiently reliable so that we can now give precise figures, like 13.8 billion years and 3.16×10^{-25} seconds, with some confidence that we know what we are talking about.

Acknowledgements

This book was originally written in Dutch and published by *Veen Magazines*. The translation into English has been carried out by Saskia Eisberg-'t Hooft and Joanne Furniss. We would like to thank them for all their effort and work, and the many questions they asked us that helped to improve the quality of the text as it is today. Words of thanks also go to our publisher, World Scientific, for their professional and enthusiastic approach to publishing this book.

We would also like to thank Christiaan Eisberg for reading the text and providing us with useful comments.

A number of people have contributed their expertise or opinions regarding specific subjects of the book. We would like to express our gratitude to Prof. Dr. J. J. Bredée for his assistance with the text about the heartbeat, to Dr. Tatiana Boiko and Dr. Elena Battaglioli for discussions about timescales in biology and specifically about the working of neurons, and to Annemarie Kleinert for the etymology of the German word *stunde*. We also thank Burchard Mansvelt Beck for his numerous remarks on the calendar and related issues enabling us to greatly improve our discussions on these themes.

Finally, we would like to acknowledge our colleagues at the Institute for Theoretical Physics in Utrecht, the Netherlands, for their keen participation in discussions, and many of our friends and family members for their encouragement and support in bringing this book to a successful conclusion.

Natural Phenomena and Their Timescales

Introduction

Time is of the essence. In natural sciences, time is an indispensable parameter. With the word 'time' we might mean to express a 'point in time', the exact moment an event takes place, or a 'time span', the period during which an event takes place. This time span could be long or short — our world is filled with such a diverse range of amazing natural phenomena, that the variances in time spans during which they take place often far exceed our imagination. On the one hand, computers nowadays are able to compute millions, sometimes billions of calculations in a second; while on the other hand, there are natural phenomena occurring on our planet that have taken millions of years to evolve. For example, the evolution of many living organisms takes place so slowly that its almost imperceptible progress is difficult for us to fathom.

But modern natural sciences show us phenomena that take things a lot farther in both directions than the two examples above. The smallest matter mankind has studied moves considerably faster than the quickest computing processes of the most expeditious machine; while on the other side of the timescale we see planets, stars and entire galaxies of unimaginably old age, some of billions of years. Scientists believe they know almost exactly how old the universe is, but even its seemingly eternal lifetime does not constitute a limit for physicists' research.

Structure of the Book

The objective of this book is to illustrate the various timescales we observe in the world around us, each of which is unique. We will start with the unit of one second — the timescale we are probably all most familiar with; the timescale that is also most widely used as a fundamental unit in modern science. From there we march onwards, with each step jumping to the next scale by a factor of 10. In the first part of our book we will look at increasingly larger timescales, examining phenomena that last for exactly 10 seconds, 100 seconds, 1000 seconds and so forth. These phenomena will be laid out and illustrated page by page, until we reach the age of the universe — and we won't stop there, as there are processes that take longer than the evolution of the universe itself!

At the other end of the scale, many events last less than a second. In the middle of this book — after we have covered the longest lasting periods that feel like eternities — we jump to the smallest units of time, to the natural phenomena that are completed in the fastest possible times. We then increase these time spans by a factor of 10 with each step, until we get to the end of the book and reach one second once again.

This may seem like a rather unusual layout, but we believe this to be the most lighthearted; this creation should really be regarded as a coffee table book, something to browse through at your leisure. Start wherever you like, hop and skip through the various pages, from one segment that piques your curiosity to the next. We would like you to discover our world as we see it: fascinating and remarkable at every conceivable timescale. Every unit of time is unique. Every level showcases enthralling phenomena. In other words, our book comprises a series of independent short portrayals and illustrations of phenomena that manifest themselves across various periods of time, from the blink of an eye to a blue moon.

Process

When we were contemplating writing this book, we did not know exactly what subjects we would come across and a lot of research was required to compile its contents. This led us to fall from one amazement into another, discovering awesome details that we are excited to share with you. For example, did you know that there are phenomena that leave a trace at virtually all time spans? A case in point is atoms and subatomic particles. These can disintegrate as a result of natural forces called 'radioactivity'. But radioactivity can be the consequence of various types of natural forces, meaning that it exhibits varying characteristics depending on its catalyst. Sometimes a particle or atom must complete a complicated process to disintegrate. In our professional jargon we refer to this as a 'tunneling process': a particle must dig a tunnel, right through a high potential barrier. This digging can take a very long time, or be over in an instant, depending on the length and depth of the particular tunnel.

Radioactive decay is described by the concept of 'half-life': the average time it takes for the number of radioactive particles of a given kind to decrease by half. This is not the same as the average lifetime of a particle. The average lifetime of a particular radioactive particle is always 1.442695 times its half-life (see graph below). At least, this is the case for particles of which the disintegration follows an exponential curve; with a few exceptions (such as the K-particle), this is always the case.

Some atomic nuclei and other particles disintegrate so quickly that it is very difficult to prove scientifically that these particles exist, while others take so long to decay that it is complicated to prove they are actually

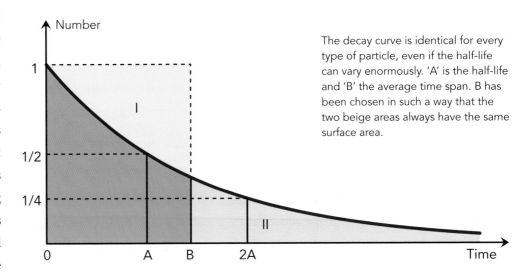

The decay curve is identical for every type of particle, even if the half-life can vary enormously. 'A' is the half-life and 'B' the average time span. B has been chosen in such a way that the two beige areas always have the same surface area.

disintegrating. We have chosen the most interesting of these particles, and those whose disintegration times, measured in seconds, also come closest to a power of 10. Information about disintegration times can be recognized by boxes with pink borders. Similarly, other recurring themes have also been color-coded.

Recurring Themes

Pink Boxes

As described above, facts and figures about decay times have been placed within pink side bars.

Blue Boxes

In the world of stars, planets, moons and comets, orbital and rotation periods prove to vary wildly. Because of their enormous variation, these times have also resulted in a recurring theme within the book. We have placed such events within a blue border.

Yellow Boxes

Periodical signals — such as electromagnetic waves — also comprise a theme. Depending on the wavelength, a frequency may vary between a mere thousand vibrations per second and the immeasurably fast vibrations of, for example, X-rays and gamma rays. We also find rhythmic and periodical systems in biology, such as our own heartbeat or menstruation cycle. Notes about these periodical signals and vibrations are placed within yellow borders.

Green Boxes

Then there is the 'new' kind of science, cosmology. Physicists and astronomers have been able to map out a large part of the universe's history. From observations and calculations they have been able to deduce that the start of the universe was an explosive one, a 'Big Bang'. Within a fraction of a second not only was all the matter that makes up our universe created, but also time and space itself. The creation of time and space still continues, meaning that our universe continues to expand. The beginning of our universe is still the subject of lots of speculation but the picture is becoming more and more clear. This is something we like to illustrate in our book as well; at a ten-billionth of a second the universe had already expanded to such a level that it enables us to use the laws of physics known at present to understand its progress. It is no mean feat to calculate back to the beginning of the universe from where we are now, but this is something we are getting better and better at nowadays. We describe the various phases of the universe within green borders.

Orange Boxes

We do not know much about the size of our universe yet. It is perfectly feasible that the universe is literally infinite. However, for the purposes of defining the universe's size, we confine ourselves to that part of the cosmos that we are able to observe with our largest telescopes. The galaxies furthest away beamed their light towards us when our universe was still quite young, about 13 billion years ago. If we consider this the boundary of our universe, then we have established a good perimeter to determine its size. After emitting the rays we are now observing, these galaxies continued to move away from us. This means that by using our own definition of size, our universe has a radius of almost 50 billion light-years (one light-year is the distance light travels within one year, or almost 10 trillion* kilometers)!

We will see, though, that the universe was quite compact and small for the first few days of its existence. It might appear strange that two galaxies could be more than 13.8 billion light-years away from one another in a universe that is only 13.8 billion years old, and within which nothing moves faster than the speed of light. This apparent paradox is explained by Einstein's Theory of Relativity. This theory states that while no matter in a given region of space can move faster than the speed of light, space-time itself does expand faster than the speed of light. This is how it is possible that there may be galaxies that are almost 100 billion light years apart. Because light and light years play an important role in this book, we refer to distances travelled by light in the various time spans discussed in the book in orange sections.

Thus, summarizing, we are using the following color schemes throughout the book to indicate the subject of the facts and figures discussed:

- decay times and half-lives
- orbital and rotation times
- periodical signals and vibration periods
- the history of the universe from the Big Bang onwards
- distances travelled by light

As mentioned, this book consists of a few hundred short, often illustrated, sections, which stand on their own. Only occasionally do we refer to other sections in the book for further details or interesting links with phenomena at other timescales. As described above, the first part of the book starts with increasing timescales. And we will not stop at the age of the universe, because chances are that the universe has a longer future than it has a past. In any case, according to modern science, there are particles that have a half-life much longer than the age of the universe. A case in point is the proton, a quite basic subatomic particle: its life span might prove to be more than a quadrillion times the current age of the universe!

But our story does not end with the lifespan of a proton — there are even longer timescales imaginable. We then venture into the world of absurdly high numbers. Our current knowledge of science is so

*We use American 'short-scale' numbers in this book. In other words, a trillion is a million million. A billion is a thousand million.

severely limited in this respect, that it is difficult to speculate about enormously large numbers. As such, we will only briefly touch upon these dark eternities.

Reverting to the concept of proton disintegration, the actual lifespan of a proton is altogether uncertain. It may be that it does not disintegrate at all — even if scientists have difficulty believing this. Most theories concerning the disintegration of the proton predict that the particle will disintegrate at some point, even if — as indicated earlier — this might take a quadrillion times as long as the age of our universe, which is also an incredibly difficult concept to comprehend. As the far future of the universe is entirely dependent on whether and when protons disintegrate, it is far from certain, which is why we will be brief on this subject too.

Borders

Natural phenomena that manifest themselves at almost unimaginably small scales are responsible for the forces that lead to the disintegration of protons. Here we are at the borders of the imaginable, where scientists can still only just conceptualize what time and space look like. It is truly remarkable that the phenomena that provide protons with such a long life span are the same phenomena with the smallest timescales. The Big Bang commences in a practically indivisible moment in time. The phenomena at play here are perhaps best explained as 'superstrings', a concept where we do not consider the most elementary building blocks of nature to be 'point-like', but vibrating 'strings', endlessly elastic, with lots

of other characteristics that are difficult to grasp. Scientists are not at all certain about string theory, which is very mathematical in nature, and not yet well understood. This part of the book is based on admittedly shaky territory, but we will not be there very long.

After superstrings we revert to longer timescales, through the various stages the universe must have passed, via the extremely short time intervals of the elementary particles – the vibrations of gamma waves, ultraviolet light, radio and sound waves, and many others — until we reach 1 second again, the end of our story. This book can be read in any direction, from cover to cover, starting at whichever end you like; from back to front if the smallest fractions of a second particularly tickle your fancy.

Measuring Time

Nowadays, clocks are extraordinarily stable, meaning time can be measured very accurately. Astronomers measure vibrations, rotations and orbits of stars and planets quite precisely, sometimes up to a millionth of a unit, and sometimes even more precise. Therefore you'll see some very exact measurements in certain parts of this book. In the most extreme cases, intervals have been measured — in seconds — with a precision of 15 decimal places, and this accuracy is continuously being improved. As such, time is the most precisely-measured quantity we have at our disposal. In second place is the measurement of length, measured in meters — with a precision of 12 decimal placess. In third place is the kilogram, with an accuracy of 10^{-8}.

Tradition

Finally, we feel duty bound to note that the setup of our book is not entirely original. Kees Boeke, a teacher in Bilthoven (close to Utrecht) in the Netherlands, published a book with the title *Cosmic View: The Universe in 40 Jumps* in 1957, which was the precursor to the well-known short film *Powers of Ten*. After Kees Boeke's project, several works have been created inspired by the theme of powers and scales. Boeke's work was based on the distance scale, meaning distances and measurements of objects. It is enormously fascinating to compare the very largest objects in our universe to the smallest ones that we have been able to examine.

We even discovered that covering time in powers of ten has been done before as well. For example, there is a film entitled *Powers of Time*, a film based on timescales, but it is not nearly as elaborate as our book. Fortunately, there is enough room for various interpretations of and perspectives on this theme. With this book we are continuing the tradition started by Kees Boeke by looking at timescales, but based on 21st-century science.

Gerard 't Hooft and Stefan Vandoren

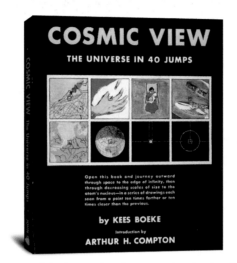

One of the last works of Kees Boeke, this book is a classic on learning about the scale of things.

Large and Small Numbers

We are used to ordering numbers in terms of powers of ten. As such, a hundred is ten times ten, or $10^2 = 10 \times 10$; a thousand is ten times a hundred, or $10^3 = 10 \times 10 \times 10$, and a million is 10^6, so a thousand times a thousand or ten times one hundred thousand.

The terminology for the higher powers of ten is not standardized worldwide. There are two scales of naming larger numbers, the *short scale* and the *long scale*. The short scale is most commonly used in English- and Arabic-speaking countries, but also in Russia. Whereas the long scale is used in most other countries in Europe, parts of South America and parts of Africa. Other countries, like China and India, employ other counting systems. The terminology used in long and short scale systems starts to differ from 10^9 onwards. The number 10^9 is called a *billion (thousand million)* in the short scale, whereas in the long scale it is called a *milliard*. A billion in long scale then stands for 10^{12}, whereas in short scale this number is called a *trillion (thousand billion)*. In the table (top right), we summarize the most important numbers for which differences occur.

The logic behind these scales is as follows. In the short scale, one counts with factors of a thousand, in the sense that a trillion is a thousand times a billion (1000×1000^3), a quadrillion is a thousand times a trillion (1000×1000^4), a quintillion is 1000×1000^5, and so forth. The long scale, on the other hand, works

Number	Short Scale	Long Scale
10^9	Billion	Milliard
10^{12}	Trillion	Billion
10^{15}	Quadrillion	Billiard
10^{18}	Quintillion	Trillion
10^{21}	Sextillion	Triliard
10^{24}	Septillion	Quadrillion

with factors of a million: a billion is a million times a million, or $1,000,000^2$, a trillion is a million times a billion, or $1,000,000^3$, and a quadrillion is $1,000,000^4$: a million times a trillion. The words milliard, billiard and triliard are not used in the short scale, and hence they are absent in modern English.

It is not difficult to find even larger numbers in nature. The number of bacterial cells on Earth is estimated to be 5×10^{30}, that is five nonillion in short scale, or five quintillion in long scale. Meanwhile, 8×10^{60} is the number of Planck-time intervals $(5.39 \times 10^{-44}$ seconds, see Chapter 22) in the lifetime of the universe, and there are about 10^{80} atoms in the observable universe.

Even larger numbers exist, which occur in problems of mathematics or computational theory. Most commonly known are the *googol*, which stands for 10^{100}, and the *googolplex*, which is 10^{googol}, or $10^{10^{100}}$.

Some large numbers are so commonplace that they have become embedded into our very language, by way of prefixes. We all know that *hecto* stands for a hundred, while a *kilo*-meter is 1,000 meters. The prefix for one million is *mega*, and for a billion (short scale) it is *giga*. We provide more unusual examples in the following table:

Prefix	Number	Name (short scale)
yotta	10^{24}	septillion
zetta	10^{21}	sextillion
exa	10^{18}	quintillion
peta	10^{15}	quadrillion
tera	10^{12}	trillion
giga	10^{9}	billion

For instance, one light-year — the distance that light travels in one year — is about 10 petameters, and the mass of the Earth is about 6,000 yottagrams, or 6 octillion grams. Similarly, there are also prefixes for small numbers:

Prefix	Number	Name (short scale)
deci	10^{-1}	tenth
centi	10^{-2}	hundredth
milli	10^{-3}	thousandth
micro	10^{-6}	millionth
nano	10^{-9}	billionth
pico	10^{-12}	trillionth
femto	10^{-15}	quadrillionth
atto	10^{-18}	quintillionth
zepto	10^{-21}	sextillionth
yocto	10^{-24}	septillionth

The fastest spinning pulsars rotate around their axis in a thousandth of a second, that is, a millisecond. The size of most atoms range between 30 and 300 picometers, and the mass of a proton at rest is about 1.6 yoctograms, or 1.6 septillionth of a gram.

The discrepancy in naming numbers within short and long scales sometimes leads to confusion, especially when translating numbers from one language to another. There is also the possibility of confusion with the terminology of the prefixes. For instance, in the words 'yotta' and 'yocto' one finds the Latin root 'octo' which stands for 'eight'. Yet, in septillion (the short scale name), we find the Latin root 'septem' which stands for 'seven'. This is confusing, since both 'yotta' and 'septillion' are used to indicate 10^{24}. (The long scale name for 10^{24} is quadrillion.) Similarly, in 'zetta' and 'zepto' we recognize 'seven', whereas sextillion clearly refers to the number 'six', though both are used to indicate 10^{21}.

Chapter 1

$10^0 = 1$
1 second

The word 'second' is derived from the Latin word *secundus* or *gradus secundus*, which means 'second step' or 'next step'. The Romans divided the daylight time into 12 hours. As a further division, an hour was first split into 60 minutes, and as a second step, each minute divided into 60 seconds.

Most mechanical clocks tick approximately once every second. The Dutch physicist, Christiaan Huygens, improved their accuracy with the introduction of the pendulum. The time that a pendulum takes to swing back and forth depends mostly on its length and not on its driving mechanism. This is why it is relatively easy to set a pendulum clock so that its hands circle the dial at a precisely defined speed.

A weight at the end of a piece of string with a length of 99 centimeters takes two seconds to swing back and forth. So the pendulum of a clock needs to be about one meter long to provide a tick — half a swing — every second. The exact length a pendulum needs to be so as to tick once per second also depends, to a lesser extent, on its shape.

The ticking of a clock reflects our instinctive human need to mark the passing of time, the fleeting moments of our lives measured out in reliable seconds. In modern science, the second is used as a fundamental unit of time. Time can be measured more precisely than any other physically observable quantity. To enable the exact measurement and definition of a second, we use the most accurate and dependable clock available to us: the atomic clock.

The Friesian clock usually ticks about once every second. It was well known for its steadfast craftsmanship and was elevated to an art form in its heyday. At the end of the 19th century, the invention of more accurate timekeepers made Friesian clocks less desirable, though in recent years these Dutch devices have regained a nostalgic popularity.

Christiaan Huygens

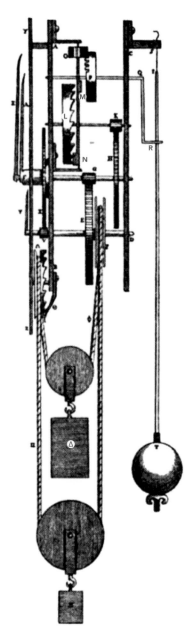

In 1658, Christiaan Huygens published this side view of his clock. The clock face is on the left side, 'MN' is the pallet verge powered by crown wheel 'L' and on the right, the pendulum moves in what resembles a fork (at 'R'). The weight Δ, which powers the clock, can be pulled up without disturbing the movement of the clockwork.

In ancient times, people believed that the rotation time of the earth around its axis was constant. So, just like the Romans, one could simply divide the average duration of a day by the number of hours, 24, and the number of seconds in an hour, 3,600, and one was left with 1 second. The rotation of the earth, however, is not quite constant. Movements in the atmosphere, the oceans and ice caps cause minimal but measurable variability. Hence the demand for a more precise definition of a second.

Nowadays, we use clocks based on the atom cesium-133 (^{133}Cs). Electrons in this atom can be made to vibrate and the frequency of this vibration turns out to be universally constant. A second is now defined as the time this atom needs to perform 9,192,631,770 vibrations. Modern atomic clocks are extremely stable and deviate no more than 1 second per 10 million years. More about this on page 148, where we talk about 10^{-10} seconds.

0.86 seconds
The average duration of a human heartbeat

The heart is a pump that provides our body with flowing, oxygenated blood. The heart of a human being at rest pumps about once a second. The average frequency is roughly 70 heartbeats per minute for men — approximately 0.86 seconds per heartbeat — and 75 per minute for women. Athletes have a somewhat slower heart rate of around 40 beats a minute, and those who do not engage in any regular exercise have a heartbeat of about 80. When distressed or intensely physically exerted, the frequency may rise to 200 beats per minute. When Neil Armstrong landed on the moon, his heartbeat was highly increased — this must have been caused by emotional tension, as he was barely able to move a muscle.

The pumping action of the human heart works as follows. In the initial phase, the heart is relaxed and both atria fill up with blood. The right atrium is provided with deoxygenated blood by the superior and inferior hollow veins, and the left atrium is provided with oxygenated blood from the lungs via the pulmonary veins.

Then, the atria contract. The blood streams into the chambers passing the valves between the atria and the chambers. Next, the chambers contract and pump the blood away from the heart. This happens by opening the aorta valve and the pulmonary artery valve. Tiny muscles attached to the valves between the atria (the tricuspid valve on the right and the mitral valve on the left) prevent leaking and the reflux of blood from the chamber to the atria.

The oxygenated blood flows through the aorta (the main artery through the body) and then through smaller arteries to all the body's organs and tissues, and the deoxygenated blood through the pulmonary artery back to the lungs, where it is provided again with oxygen.

The cycle is now complete and another heartbeat brings the next cycle into motion. The complex action takes less than a second on average.

The natural pacemaker of the heart, the sinus node, controls the basal heart frequency. It consists of specialized muscle tissue in the wall of the right atrium. The sinus node periodically fires electrical impulses that force the heart muscles to contract. With physical exertion or stress, muscles and other organs require more oxygen, which is delivered by blood. When that happens, the sinus node makes the atria and chambers contract quicker, causing the heart to beat faster.

0.5 mV

0.3 seconds

Cross-section of a human heart. On the left is an electrocardiogram which shows the electrical activity with every heartbeat.

3

0.995 seconds
The half-life of ^{79}Zn

Isotopes are atomic nuclei that contain the same number of protons as the element they are related to, but differing numbers of neutrons. Stable isotopes exist practically forever, unlike unstable isotopes, whose half-lives vary enormously. It so happens that the half-life of one element, called ^{79}Zn, is almost exactly 1 second.

The weight of an atom is comprised of the sum of the weight of its protons and neutrons. The mass number, the sum of the number of protons and neutrons, is usually indicated to the left of and above the abbreviation of the element in question, just as we have done in this book (for example: ^{79}Zn). The 'atomic number' only counts the protons, often indicated in scientific literature to the bottom left of the name of the element (for example: $_{30}$Zn).

Since neutrons have no charge, the electric charge of a nucleus is derived only from its protons and is thus defined by the atomic number. The number of electrons circling the nucleus is equal to the atomic number as well, since atoms are usually electrically neutral and electrons and protons have equal but opposite electric charge. This explains why an isotope usually has almost the same chemical characteristics as the element it is derived from. Therefore, chemically, zinc invariably behaves as zinc, irrespective of the number of neutrons in its nucleus.

Zinc is a metal with atomic nuclei that contain exactly 30 protons each. On average, 30 electrons circle each nucleus, as this number of negatively-charged electrons neutralizes the positive charge of its 30 protons. The stable isotopes (variants) of the element zinc contain 36, 37 or 38 neutrons. However, there are dozens of unstable isotopes with either a larger or smaller number of neutrons. Zinc-79 (^{79}Zn), for example, contains 49 neutrons.

The nucleus spins around its axis very quickly: its spin, or more precisely its angular momentum, is 9/2 in natural units, while the normal spin for a zinc nucleus is 0 or 5/2. The isotope ^{79}Zn can emerge from nuclear reactions, only to fall apart after about a second by emitting an electron. By losing an electron the element disintegrates into another element, gallium, which has 31 protons and 48 neutrons. In some instances, another neutron escapes.

On page 15, at 10^3 seconds, we look at neutrons and protons in more detail, including the matter of isotopes.

1 second

On planet Earth, this is the time it takes for a stone — or any other heavy compact object for which air resistance can be ignored — to fall to the ground from a height of 5 meters. On the moon we would have to drop the same stone from only 81 centimeters for it to reach the ground in 1 second.

The Earth and the Moon, photographed from the Galileo spacecraft.

1 light-second

In an empty region of space, light travels 299,792,458 meters in 1 second — in other words, the distance of 1 light-second. The terms 'light-years' or 'light-seconds' are not measures of time but of distance. Nowadays, the speed of light is measured so accurately, that we define 1 meter as the distance that light travels in 1/299,792,458th of a second. Elsewhere in this book we will compare the distance that light travels during other time intervals.

1.28 seconds

This is the time it takes for a light or radio signal to travel the distance between the Earth and the Moon. On average — from center of gravity to center of gravity, and taking into account the elliptical shape of the lunar orbit — this is a distance of 384,403 kilometers. The speed of light is 299,792,458 meters per second, meaning that our light shines on the Moon within 1.28 seconds. Having a phone conversation with an astronaut on the Moon means you have to wait about 2.5 seconds for a response — the time it takes for a radio signal to travel back and forth.

No signal will ever travel faster than the speed of light, which means that even with extremely advanced techniques in the far future we will never be able to shorten this communication time, but perhaps there will be ways to lessen the annoyance factor of these waiting periods.

Chapter 2

10¹ = 10
10 seconds

A time span of 10 seconds is also known as a decasecond. Deca is derived from the Greek word *deka*, which means 10. The prefix deci, on the other hand, comes from the Greek word *decimus*, which means 'a tenth'. In the official numbering system the words deca, deci and hecto have been eliminated — only powers of 1,000 are named.

9.58 seconds
World record, men's 100 meter sprint (2014)

On July 6, 1912, the American Don Lippincott ran the 100 meters in 10.6 seconds. It took 50 years for an athlete to run the same distance in exactly 10 seconds. The West German athlete, Armin Hary, was the first to achieve this milestone in 1960. A relatively short time later, in 1968, the 10-second barrier was breached for the first time, with a new world record of 9.95 seconds, achieved by another American, Jim Hines.

In August 2009, the world record was set at 9.58 seconds. In other words, a full second was won within a century — about a tenth of a second per decade. This continuing improvement is attributed to progress in the training, nutrition and condition of the athletes.

The fastest man in the world, Jamaican Usain Bolt, became the world record holder of the 100m in 2009 with a time of 9.58 seconds. The athlete is very tall for a sprinter at 1.96 meters. With such long legs, he is not the fastest out of the starting blocks, but once he reaches full speed, his paces measure 2.44 meters. Running a distance of 100m, he needs three to four fewer paces than his competitors — 41 steps in total in his fastest race. His maximum speed so far was clocked at 44.72 kilometers per hour.

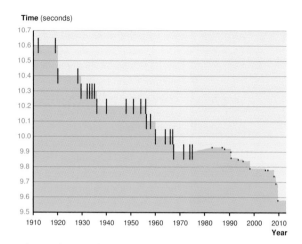

Time (seconds)

The evolution of world record speeds for the men's 100m sprint. After 1976, electronic sensor systems (FAT: Fully Automatic Timing) and the photo finish, which is able to measure intervals of 1/100th of a second, greatly improved precision.

10.18 seconds

The time it takes to free fall from a height of 508m

The Taipei 101 tower in Taiwan is 508 meters high. If we ignore air resistance, any object dropped from the roof of this tower will take 10.18 seconds to reach the ground — irrespective of its size or mass. At the moment of impact, the object will have reached a velocity of 99.83 meters per second, or 359.40 kilometers per hour. To perform a free fall of exactly 10 seconds, you would have to start from a height of 490.5 meters.

If we do take air friction into account, then the time it takes to free fall is considerably longer. The amount of friction depends on the shape and size of a falling object. To give an example, close to the surface of the earth, the human body cannot attain a higher

Galileo Galilei

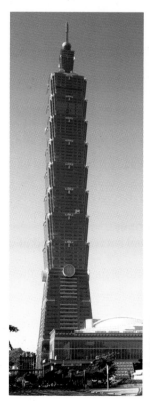

The Taipei 101 tower   The Leaning Tower of Pisa

Galileo Galilei (1564–1642), an Italian physicist from Pisa, discovered that all objects, in the absence of air resistance, fall with the same acceleration irrespective of their shape or weight. In the gravitational field of the Earth, this acceleration is 9.80665 meters per second squared (m/s²). In other words, with each second, an object gains velocity by almost 10 meters per second.

Legend has it, according to Galileo's disciple Vincenzo Viviani, that the physicist made his discovery by dropping two different balls — one weighing ten pounds and another just one pound — from the upper ledge of the Leaning Tower of Pisa. The tower, however, is only 56 meters high, meaning that the balls would have taken just 3.38 seconds to hit the ground. It would have been virtually impossible to determine that the two objects reached the ground at exactly the same time using only the naked eye — if there had been any distance between the two, it would have been minimal. Galileo did, we believe, also suggest letting the balls roll down a slope to enable relatively easier time measurements.

In any case, it is unlikely that Galileo ever did these kinds of experiments himself; he probably deduced the outcome from theoretical arguments instead of experimentation. In fact, experiments with falling objects had been carried out before by others, such as Simon Stevin in 1586, and the outcome always confirmed that Aristotle's old theory was untenable; Aristotle believed that heavy objects fall faster than lighter ones, and that all objects maintain their speed.

velocity than 70 meters per second — although that is still more than 250 kilometers per hour. At this rate, the air's friction compensates for the Earth's gravity, and velocity ceases to increase. From that moment onwards, there is no more acceleration. But further away from the Earth's surface, in higher layers of the atmosphere, friction is smaller and higher speeds can be reached during free falls.

10.2 seconds
Half-lives of ^{144}Eu and ^{75}Zn

The half-lives of various radioactive materials vary widely. There are many nuclei that exist only for a few seconds. We have chosen two whose half-lives are near the 10-second mark.

About two parts in a million in the Earth's crust consist of the heavy metal europium. The element's atomic nucleus contains 63 protons and it is stable if it contains approximately 97 neutrons. Europium-144 (^{144}Eu), however, contains only 81 neutrons. It can emerge from a nuclear reaction or after radioactive disintegration of ^{144}Gd — an isotope of gadolinium.

With a half-life of approximately 10.2 seconds, ^{144}Eu disintegrates as a result of 'electron capture'. This occurs when one of its own electrons comes too close to the nucleus. The electron is absorbed, causing the nucleus to transmute into samarium-144 (^{144}Sm, with 62 protons and 82 neutrons). Europium has important applications, such as for televisions and computers, where it can create bright red colors.

Sirius A, the brightest star in our sky, with the white dwarf Sirius B, the small dot in the bottom left hand corner. The light in this picture, taken by the Hubble Telescope, has been artificially enhanced to make the white dwarf visible.

Sirius B has a diameter of about 12,000 kilometers. That is about average for a white dwarf, most of which are around the same size as our planet Earth. Sirius B's mass, however, is approximately the same as our Sun's. The diameter of the Sun and Sirius A, which are normal stars, is much bigger, approximately 1.4 million kilometers for the Sun and slightly larger for Sirius A. Sirius A's mass is approximately twice that of our Sun.

The zinc isotope ^{75}Zn also has a half-life of 10.2 seconds. It behaves just like ^{79}Zn (see page 4). ^{75}Zn, however, has a spin of 7/2 instead of 9/2.

20 seconds
Minimal rotation period of a white dwarf

Stars and planets almost always rotate about a fixed axis. They have this in common with both the largest objects in the universe as well as the smallest. The velocity with which a star or planet can spin is limited. Planets consisting of ice or rock will never be able to complete a rotation in less than one hour, otherwise they would burst into pieces. Only if they consist of much denser matter, are they able to spin faster.

White dwarfs are burned-out stars at the ends of their lives. Most of them have roughly the same mass as the Sun, but the same size as the Earth. As such, the mass density of a white dwarf may be as much as hundreds of tons per cubic centimeter. Only neutron stars and black holes have higher densities.

In the nucleus of a white dwarf, no nuclear reactions occur. All hydrogen and helium have been converted into carbon, oxygen and heavier elements, such as iron. This is why it is difficult to spot a white dwarf; it emits hardly any light. Sirius B, which forms a double star together with Sirius A, is the most well-known and visible one.

Some white dwarfs are so close to a companion star that they suck up matter; this is because they are so heavy, their gravity is much stronger than that of the companion star. This process results in an increased angular momentum, meaning that the white dwarf will turn around its axis faster and faster until it reaches the critical velocity of 1 revolution per 20 seconds. After that, the white dwarf itself loses matter by emitting electromagnetic waves, after which it loses rotational velocity again (see picture below). Calculations have determined that the maximum rotational velocity is 1 rotation per 20 seconds. Only pulsars — also star-like objects — rotate around their axes much quicker (see 10^{-3}, 10^{-2}, 10^{-1} and 1 second, in Chapters 45–48, at the end of the book).

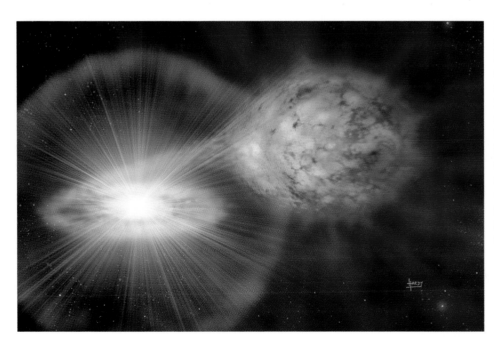

Drawing of a binary system with a white dwarf. On the right, a red giant known to be a 'swollen' star is emitting a lot of hydrogen at the end of its life. A fraction of the emitted hydrogen is taken in by the neighboring white dwarf, causing it to rotate quicker. If this process continues long enough, the red giant turns into a white dwarf, and the white dwarf's mass grows too much. This can result in an enormous explosion, potentially creating a neutron star or even a black hole. Sometimes, such explosions are visible to the naked eye here on Earth.

Chapter 3

$10^2 = 100$
100 seconds = 1 minute, 40 seconds

100 seconds are also called one hectosecond. The prefix 'hecto' to indicate 100 is not common anymore. It is derived from the Greek *hekaton*, which means 'a hundred'. The word 'minute' comes from the Latin *pars minuta prima* or 'the first small part'. This signifies the first small part of an hour or, more specifically, 1/60th. The second small part is then the second, as described in Chapter 1.

The division of the hour into 60 minutes can probably be ascribed to the Babylonians. Babylon was situated in what used to be Mesopotamia, currently better known as Iraq. The Babylonians — or maybe even their predecessors, the Sumerians — introduced the sexagesimal (base-60) system more than 5,000 years ago. Not only did they divide the hour into 60 minutes and then 60 seconds, but degrees in geometry and astronomy were also divided into 60 arcminutes, and every arcminute into 60 arcseconds.

A circle, then, encompasses an angle of 6×60 = 360 degrees. The fact that sexagesimal systems were introduced is possibly explained by the fact that the number 60 has so many divisors: 2×30, 3×20, 4×15, 5×12 and 6×10. The Romans already used the number 10 as the basis of their counting system, but the current decimal (base-10) system, where fractions can be written with numbers ranging from 0 to 9 after a period (such as 0.5 for a half), was introduced much later by the mathematician Simon Stevin.

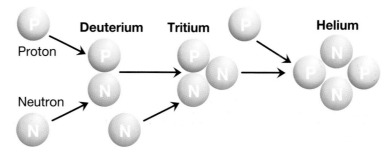

Nucleosynthesis. While the universe is rapidly cooling down, the first atomic nuclei are created through nuclear fusion.

100 seconds after the Big Bang

When it was just 100 seconds old, the universe was already 80 light-years in size. A light-year is the distance that light travels in a year, which is about 10 trillion kilometers. The average density of the universe had decreased to 1 kilogram per cubic centimeter. In this short time, the universe had already traversed several phases — as described in the second part of this book.

At 1 second after the Big Bang, the temperature was still 10^{10}K, which is 10 billion kelvin (1K is 1 kelvin and a temperature of 0 kelvin — also called absolute zero — equals 273.15 degrees Celsius below 0. The difference in temperature expressed in kelvin or Celsius is thus 273.15 degrees). The universe consisted of protons, neutrons, electrons and photons that were moving in all directions. Particles such as neutrinos and unknown dark matter existed as well, but did not interact; they contributed to gravity, crucial to the evolution of the universe.

Then, about 3 minutes after the Big Bang, the process of nucleosynthesis commenced. This caused protons and neutrons to cluster together in groups of 2, 3 or 4 particles. This is how, in addition to protons and neutrons, the atomic nuclei of the elements deuterium (heavy hydrogen), lithium and helium emerged. The temperature dropped further to 10^9K.

The relative quantities of photons, protons, deuterium and other nuclei that developed at this point can be calculated very precisely, since their interactions have been studied in laboratories. Interstellar gas clouds exist, wherein the ratios of these elements have changed very little since shortly after the Big Bang — this means that these clouds can be used to test ideas regarding the conditions at the beginning of the universe.

It has been determined that nucleons — protons and neutrons — quickly began to convert into helium. Within only a few minutes, approximately 25% of the nucleons transmuted into helium, while a much smaller fraction — about 20 parts per million — convert into deuterium. Heavier atomic nuclei could not have been created that fast or in a similar way, because stable nuclei with 5 or 8 nucleons do not exist (therefore, this prevented the building up of heavier elements by adding nucleons one at the time). Only much later were the nuclei of heavier elements created, within the centers of heavy stars.

70.60 seconds
The half-life of ^{14}O

95.6 seconds
The half-life of ^{74}Zn

122.24 seconds
The half-life of ^{15}O

There are three stable isotopes of oxygen: ^{16}O, ^{17}O and ^{18}O. Their nuclei contain 8 protons and 8, 9 or 10 neutrons. But isotopes ^{14}O and ^{15}O have only 6 or 7 neutrons, making them unstable with a lifetime of around 100 seconds. These isotopes come into existence after nuclear reactions resulting from collisions between other nuclei, or after radioactive decay of other nuclei.

Both types can decay, either by emitting a positron (which is a positively charged electron) or by capturing an electron from one of the inner electron shells of the atom. This causes a proton in the nucleus to change into a neutron, creating a stable nitrogen isotope — both ^{14}N and ^{15}N are stable.

The metal element zinc, described earlier, also has an isotope that has a half-life near the 100 second mark: ^{74}Zn has a half-life of 95.6 seconds.

100 seconds

Chopin's *Minute Waltz*

In 1847 Frédéric Chopin created the *Minute Waltz*, a musical piece for the piano. Depending on how fast the piece is played it will take the pianist between 1.5 to 2.5 minutes to finish. The Waltz is often played

Frédéric Chopin (1810–1849)

extremely fast, in tempo *molto vivace* — sometimes more than 160 clicks per minute on the metronome. Its motif has many modern applications; some mobile phones have its melody as their ringtones.

The official name of the piece is Waltz Op. 64, No. 1 Minute in D flat major, but Chopin initially called it *Le Petit Chien*. Chopin envisioned a little dog trying to catch its tail following the rhythm of the musical notes. Below is the opening of the *Minute Waltz*.

Chopin preferred writing musical solo pieces for the piano, such as this *Minute Waltz*, but when he was younger he also tried his hand at composing pieces for piano and concert. In these, however, the concert instruments often played second fiddle to the piano.

101.04 seconds

World record men's 1,500 meters speed skating on ice

On 9 November 2007, Dutchman Erben Wennemars set the world record for the 1,500 meters speed skating on ice at 1 minute, 42 seconds and 32/100th of a second. This record was improved by just 21/100th of a second by the Canadian Denny Morrison on 14 March 2008. The following year, the American Shani Davis broke the world record — twice. On 11 December 2009, he set a new time of 1 minute, 41 seconds and 4/100th of a second. His average speed was calculated at 53.444 kilometers per hour. At the subsequent Olympic Games in 2010, Dutchman Mark Tuitert beat Davis to take the gold medal with a time of 1 minute, 45 seconds and 57/100th of a second. But at the time of writing this book, Shani Davis was still the world record holder.

The women's world record for the same distance has been held by the Canadian Cindy Klassen since 2005. Her time was 1 minute, 51 seconds and 79/100th of a second. The closest anyone ever came to skating the 1,500 meters this fast was the Dutch woman Irene Wust, with a time of 1 minute, 52 seconds and 38/100th of a second, in March 2007.

The 1,500 meters speed skating is often referred to as the King's Distance, the most difficult discipline on ice. This is because the distance is extraordinarily taxing and skaters often experience intense pain and acidification of the muscles. The 1,500 meters is

American Shani Davis, world record holder of the 1,500 meters speed skating on ice.

too long for sprinters, who lose time in the last two rounds, but too short for long-distance skaters, who cannot bring about a fast and explosive enough start. As both long- and short-distance skaters compete at this distance, the 1,500-meter race is usually very exciting and unpredictable.

102 seconds
World record men's 200m freestyle

This world record-breaking time was swum in July 2009 by the German Paul Biedermann. The crown was previously worn by Michael Phelps, who won eight gold medals at the 2008 Olympic Games in Beijing.

103 seconds
The half-life of ^{231}Ra

Radium has atomic number 88, which means its nucleus consists of 88 protons. This element was discovered by Pierre and Marie Curie in 1898. This triggered further research into radioactive materials. The most stable isotope of radium is ^{226}Ra, with a half-life of 1,600 years. An extra 5 neutrons takes us to isotope ^{231}Ra, which has 143 neutrons in its nucleus. Via beta decay, a type of radioactive process in which a beta particle (in this case an electron, but sometimes a positron) is emitted from an atom, actinium is created — or to be more precise, the isotope ^{231}Ac. As we see elsewhere in this book, beta decay causes a neutron to be converted into a proton, electron and an anti-neutrino (see also page 15). The atomic number of actinium is thus 89.

Michael Phelps

Marie Curie-Sklodowska (1867–1934). Together with her husband Pierre, she discovered the extremely radioactive element radium. She won two Nobel Prizes, the first in 1903 — together with her husband and the French physicist Henri Becquerel — for physics and the second in 1911 for chemistry, for the discovery of the two elements radium and polonium (named after Poland, Marie Curie's home country). Madame Curie was the first woman to receive a Nobel Prize and the first person to receive two Nobel Prizes. She died in 1934 from the effects of prolonged exposure to high doses of radiation during her career.

Chapter 4

10^3 = 1,000
1,000 seconds = 16 minutes, 40 seconds

1,000 seconds makes 1 kilosecond — that is a little over 15 minutes. At some universities, mainly in Europe, this is known as the 'academic quarter': this stems from the time when the ringing of the church bells was the general method of timekeeping and when the bells rang, you had 15 minutes to get to class.

In modern times, this academic quarter is often interpreted by students to mean that if the lecturer has not shown up within that time, the class has been cancelled and they get the morning or afternoon off — often to the chagrin of the professor, who might simply have found it difficult to get to his class on time.

885.7 seconds = 14.76 minutes
The average lifetime of a free neutron

The neutron is a subatomic particle without an electric charge. We usually find it in the atomic nucleus, but large quantities of free neutrons can also be found in a nuclear reactor, where they sustain the fission process. Also, they are released in large quantities in the event of a nuclear explosion.

The average lifespan of a free neutron is 14.76 minutes and its half-life is 613 seconds — about 10 minutes. A neutron decays into a proton, electron and an anti-neutrino.

Neutrons and protons together form the nucleus of an atom. Electrons circle the nucleus. The decay of a neutron in the nucleus is one of the basic principles of radioactivity. But the decay

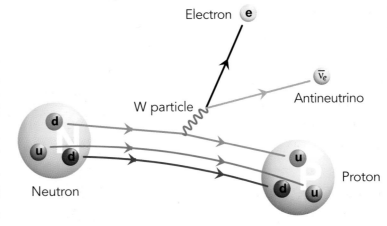

time of bound neutrons can be much longer or shorter than that of free neutrons. This is why you will find such enormous ranges of half-lives in isotopes, which differ from each other in the number of neutrons in their nuclei.

The disintegration of a neutron, also referred to as beta decay. The neutron was discovered by the British physicist James Chadwick in 1932, who received the Nobel Prize for Physics for this discovery in 1935. Chadwick also determined the mass of the neutron; the most precise measurements today indicate a mass of $1.6749274 \times 10^{-27}$ kilograms. As such, the neutron is a little heavier than the proton.

The process by which a neutron converts into a proton is called the weak nuclear force and is one of the fundamental forces of nature. Forces are transferred by exchanging special particles, in the case at hand, by W particles. More about these in Chapter 23.

Neutrinos and their antiparticles, antineutrinos, are fundamental particles with a very small mass. They do not have an electrical charge, which means they are difficult to detect. Neutrinos are detected in cosmic radiation, although most of them go right through everything.

The neutron is not a point-like particle, but consists of 3 quarks (that is, 1 up quark and 2 down quarks). The proton is very similar to a neutron, but has a positive charge and consists of 2 up quarks and 1 down quark. We discuss quarks in more detail in Chapters 23 to 25.

998 seconds = 0.998 × 10³ seconds = 16 minutes and 38 seconds
A ray of light travels from the Sun to the Earth and back

The distance between the Earth and the Sun is approximately 150 million kilometers — an average distance of 149,598,000,000 meters to be exact; this number is an average, because the Earth's orbit around the Sun is elliptical, not circular. The orbit of the Earth is not constant in time either,

as the presence of other planets influences its orbit. On average, the distance between the Earth and the Sun diminishes by 80 meters every year. Nevertheless, the distance is of such importance in astronomy that it is used as a length unit — the Astronomical Unit (AU). For instance, the distance between Jupiter and the Sun is 5.2 AU. Pluto is 40 AU from the Sun.

The speed of light is 299,792,458 meters per second. Thus, it takes a ray of light 8 minutes and 19 seconds to travel from the Sun to the Earth. Back and forth, this takes 16 minutes and 38 seconds.

Earth and the Sun

1,026 seconds = 1.026×10^3 seconds = 17.1 minutes

The fastest time on the Four Miles of Groningen

A time of 17.1 minutes, or 17 minutes and 6 seconds, is the record set at the Four Miles of Groningen race in 2011 by the Kenyan athlete, Vincent Yator. Four miles is 6,437 meters. That means Yator sustained an average speed of 22.58 kilometers per hour. The Four Miles of Groningen is one of the Netherlands' biggest running competitions, held in the north of the country.

1,060.80 seconds = 17.68 minutes
The half-life of ^{80}Br

The name of the element bromine is derived from the Greek bromo, which means 'stench'. That the element received this name is understandable, because bromine emits a foul odor at room temperature and under normal pressure. There are 2 stable isotopes of bromine that exist naturally on Earth, ^{79}Br and ^{81}Br. The atomic number of bromine is 35, as there are 35 protons in its nucleus. Bromine is poisonous and dangerous to the environment and human health. The isotope ^{80}Br is unstable. After a little over 15 minutes, it decays into the colorless noble gas krypton (Kr).

Vincent Yator at the Four Miles of Groningen race.

1,339 seconds = 22.32 minutes
**The world record for static apnea —
holding your breath**

The Guinness world record for holding one's breath is currently held by the Croatian freediver, Goran Čolak. On 29 September 2013, Čolak took the crown at a static apnea event in Zagreb, Croatia with a time of 22 minutes and 32 seconds.

The diver is allowed to inhale pure oxygen before starting his attempt, to ensure a maximum of stored O_2. This is formally called 'oxygen-assisted static apnea': the medical term for temporary respiratory arrest — not breathing — is apnea. The record is always performed under water, because it is easier that way to ensure no one cheats.

The world record for holding one's breath without inhalation of pure oxygen was set at 11 minutes and 35 seconds in June 2009. Without practice, most people can only hold their breath for about a minute or two. This is because the inhaled oxygen (O_2) is converted into carbon dioxide (CO_2). When you stop breathing, CO_2 cannot exit the lungs, and concentrations of CO_2 build up. This causes acidity to increase, which leads to reflexes in the breathing muscles and the necessity to breathe. Without it, blood to the brain does not contain enough oxygen and you suffocate. To enable you to hold your breath longer, it is essential to store as much oxygen in the lungs as possible and to learn to tolerate the pain that is caused by the reflexes, instigated by the increasing levels of CO_2.

David Blaine, one of the previous record holders for static apnea, at his record attempt in 2008. He is assisted into the water, while breathing extra oxygen (see left), trying to store as much oxygen as possible in his lungs. We see Blaine's breath-holding attempt at the moment that his oxygen is taken away and the clock starts (see top).

Chapter 5

$10^4 = 10,000$
10,000 seconds = 2.78 hours

10,000 seconds, or almost 3 hours, is the duration of many concerts, films and other spectacles. It is about the maximum amount of time anyone would like to spend on any single activity. You are often allowed 3 hours to complete a university exam. In our Western society it is unacceptable to be 3 hours late for an appointment, but in many countries this is actually quite common.

3,600 seconds = 0.36×10^4 seconds = 1 hour

The word 'hour' stems from the Latin *hora* and the Greek *ora*. The Greeks borrowed this word from the Babylonians, who divided a day — including the night — into 12 hours, meaning that their hour lasted twice as long as ours. The reason for dividing the day into 12 hours is believed to stem from the fact that there are 12 months in a year. Counting in ancient civilizations employed a duodecimal (base-12) system based on the phalanx bones of the fingers: it is easy to count the three phalanx bones on each of the four fingers using your thumb.

Although they borrowed the Babylonians' lingo, the Greeks decided instead to divide daylight time into 12 parts, calling them hours. The Romans also adopted this system. After that, the night was divided into 12 parts as well. The drawback of this system was that the actual length of the hours would vary greatly, as the duration of daylight in summer is much longer than in winter. This is why, later on, the hour was redefined as 1/24th of an average solar day — the time between two sunrises. These fluctuate as well though, as explained in Chapter 1.

The Germans use the word *stunde* to indicate an hour. It is derived from the old High German *stunta*, which means 'to stand a little'. In plural *die stunden* refers to the 12 tutelary goddesses of the times of the day in Greek mythology — from Auge, the goddess of the first light, to Hesperis, the goddess of the start of evening, and Arktos, the goddess of the night sky.

1.6 hours and longer
Orbital periods of artificial satellites

At present, more than 2,000 satellites circle the Earth at various heights, with varying orbital periods. The first satellite, launched in 1957 by the

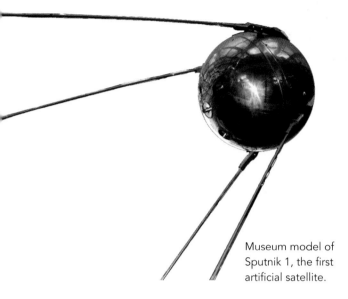

Museum model of Sputnik 1, the first artificial satellite.

height above the ground must amount to exactly 35,786 kilometers.

These days, satellite and space programs are increasingly reliant on international collaboration such as those between the American NASA (National Aeronautics and Space Administration) and its European counterpart, ESA (European Space Agency). Such an alliance gave us the International Space Station (ISS), which has been permanently occupied since November 2000. The ISS is located approximately 340 kilometers above the Earth and has an orbit of 91.6 minutes or 5,496 seconds. In its present shape, its dimensions are 52 meters in length, 82 meters in breadth and 27.4 meters in height (see picture on page 21).

9,036 seconds = 0.9 × 10^4 seconds = 2.51 hours
The half-life of ^{65}Ni

Nickel has atomic number 28, meaning it has 28 protons in its nucleus. It is a silvery-white metal. The most common type of nickel is ^{58}Ni. However, the most stable type is actually ^{62}Ni, which boasts 34 neutrons that give it the highest amount of binding energy per nucleon of any atomic nucleus – even more than iron-56 (^{56}Fe). It is believed that much of the natural nickel found on Earth originally came from meteorites.

By contrast, ^{65}Ni has 37 neutrons and is an isotope with a half-life of 2.517 hours or 2 hours and

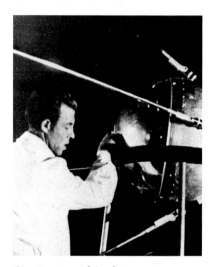

Construction of the first Russian artificial satellite, which was launched on 4 October 1957.

former Soviet Union, was called Sputnik 1. This satellite had an orbital period of 96.2 minutes (1.6 hours). Sputnik 1 hovered about 223 to 950 kilometers above sea level, high up in the atmosphere, and its orbital velocity relative to the Earth was 29,000 kilometers per hour. Lots of satellites have a comparable orbital period but not all, as their altitude and orbit may differ.

In 1974 the Dutch Astronomical Netherlands Satellite (ANS) was launched, with an orbital period of 98.2 minutes. There are also satellites whose orbital periods are exactly one natural day. These so-called geostationary satellites rotate synchronously with the Earth, which means that from our perspective they appear to remain at the same location in the sky above the equator. The

The International Space Station (ISS)

31 minutes. The nucleus of this atom has a spin of 5/2 and it decays into copper, ^{65}Cu. Copper has atomic number 29, which is 1 proton more than nickel. Again, when ^{65}Ni breaks down, a neutron is converted into a proton through radioactive beta decay. Because of its short lifetime, ^{65}Ni is not an isotope that exists naturally. It is produced in cyclotrons, by exposing copper or nickel to radiation, and in nuclear reactors, where it is produced as a waste material of uranium nuclear fission.

^{65}Ni has special applications in scientific research — for example, for the identification of minute traces of nickel in samples with concentrations of less than 1 microgram.

9,900 seconds = 0.99 × 10⁴ seconds = 2.75 hours
The half-life of ^{93}Tc

There are approximately 30 known isotopes of the element technetium, from ^{85}Tc to ^{115}Tc, with half-lives varying from microseconds (for example, ^{113}Tc) to millions of years (for example, ^{98}Tc). None of the isotopes is stable. The atomic number of technetium is 43.

Natural technetium is almost non-existent. It is created as a fission product of heavy nuclei such as uranium-238. At temperatures of less than −262°C, technetium becomes a good superconductor. ^{93}Tc has a half-life of almost 10,000 seconds, after which it decays into molybdenum. Technetium has many applications in medical diagnostics, as it emits beta rays (electrons) without companion gamma rays (intensive photons).

10,800 seconds = 3 hours
Blooming time of the Flower-of-an-Hour (*Hibiscus trionum*)

As its name would suggest, the flowers of this hibiscus plant bloom for just a few short hours before they die off. Although the rather fussy plant blooms profusely, producing fresh petals each day, it flowers only in the morning and, it is said, only when it is sunny. Nevertheless, the plant thrives in wet soil throughout Southern and Eastern Europe, and also in large parts of Asia and Africa.

The Flower-of-an-Hour (*Hibiscus trionum*). This plant only blooms for 3 hours.

10,800 seconds = 3 hours
The duration of a flight by Concorde from London or Paris to New York

The supersonic passenger jet Concorde flew from Europe to New York in less than three hours. The fleet was decommissioned in 2003, after financial trouble and a crash that killed 113 people. Concorde was capable of flying at twice the speed of sound, or Mach 2.

The speed of sound depends on the temperature and the medium through which the sound waves propagate. In air with zero humidity, and with temperatures T of up to 15°C, the speed of sound v is expressed as $v = 331.3 + 0.6\,T$, in meters per second. For example, at 0°C, Mach 2 is 2,385 kilometers per hour. At a height of 18 kilometers, where Concorde normally cruised, the temperature is much lower, around −50°C, and so is the speed of sound. Applying the formula at this temperature shows that Mach 2 = 2,169.36 kilometers per hour. The pressure of the air does not play a role.

Concorde's fastest measured speed was 1,920 kilometers per hour, but its maximum speed was

The Concorde, with its beautiful streamlined shape and striking delta wings. The nose of the plane could be pointed downward to give pilots a better view of the landing strip. Plans exist in France and Japan to create a new supersonic aircraft that flies from Tokyo to New York in only 6 hours.

over 2,000 kilometers per hour. In the 1960s, the Russians unveiled an even faster supersonic airplane, the Tupolev Tu-144, which had a maximum speed of 2.35 Mach.

3.3 hours = 3 hours and 18 minutes
The rotation period of an object around a sphere with the density of water, under the influence of gravity

The laws of gravity according to Newton describe the elliptical orbits of moons and planets around a heavy mass such as the Sun. The rotation period squared is proportional to the average distance to the center of the sphere to the third power, and inversely proportional to the sphere's total mass. If the object's orbit comes close to the surface of the sphere, then only the *density* of the sphere determines the rotation period of the object. If the sphere's density is the same as water, the rotation period of the object will be 3 hours and 18 minutes. If the sphere's density is higher, the rotation period will be shorter; when the density is lower, or the distance from the sphere's surface bigger, the period will be longer.

The Earth has an average density of about 5.5 times that of water, meaning that the rotation period of satellites is a bit faster than 3 hours. Saturn is much bigger than the Earth, but less dense, and slightly less dense than water; the rotation period of Saturn's rings is therefore a little over 10 hours. If you were to conduct an experiment in zero external gravity with two metal balls a few centimeters in size, circling under the influence of each other's gravity, their rotation periods would be approximately 1 hour.

The rotation period of 3 hours and 18 minutes is a fundamental gravitational characteristic of water.

15,900 seconds = 4 hours and 25 minutes
Signal to Pluto

The American spacecraft New Horizons is making a voyage to the dwarf planet Pluto, which will take 9.5 years. In its hedgehop past Pluto, detailed recordings of the planet will be made from a distance of only 10,000 kilometers. The radio signal with all scientific data from these recordings will take 4 hours and 25 minutes to cross the enormous distance between Pluto and the Earth. Instructions from Earth to the spacecraft take just as long. The signal from the spacecraft will be very weak at that distance: about 600 bps.

Impression of the spacecraft New Horizons, which is due to arrive at Pluto in 2015.

Chapter 6

10^5 = 100,000
10^5 seconds = 1.16 days = 27.78 hours

This chapter covers one of the periods in time which is most familiar to us, that is 100,000 seconds or 27.78 hours: a little longer than a full day — 1 day and 4 hours to be precise.

There are two types of day: the sidereal or star-based day, and the synodic or solar day. The solar, sun-based day is what we call a normal day — the time it takes for the Earth to rotate around its axis, as observed from the Sun. In other words, the period between two sunrises. This day lasts 24 hours by definition, because of the original meaning of the word hour.

A star-based day lasts 23 hours, 56 minutes and 4.09 seconds. This is the time it takes for the Earth to rotate around its axis, seen from a faraway star. These two days are not the same, because the Earth rotates around the Sun. This is further explained in the figure on the right. The sun-based day will always be a bit longer than a star-based day. For planet Earth, the difference is just over 4 minutes — for other planets, such as Mercury, the difference could be much longer (see Chapter 9).

But hold on a second! An hour is defined as 1/24th of a synodic day, the minute as 1/60th of an hour and a second as 1/60th of a minute. The question is, whether this second is the same as the second

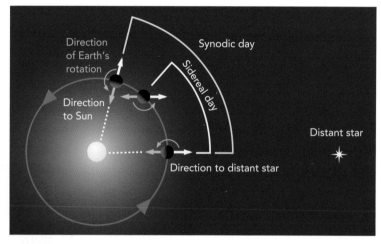

The difference between a sidereal and synodic day (not to scale).

described in Chapter 1, which was based on atomic clocks. It appears this is not the case. There is a tiny, one might joke 'minute', measurable difference.

Based on an atomic second and a day defined as 24×60×60 = 86,400 atomic seconds, then the synodic day lasts approximately 24 hours and 0.002 atomic seconds, 2/1000th of a second longer, because the rotation period of the Earth is not entirely stable. This also means the 0.002 is an average. If fluctuations in the rotation period of the Earth's axis create a difference between solar time and atomic time of more than 1 second, then atomic clocks are adjusted to compensate for this effect. This is called a 'leap second'. Since 1972, 25 leap seconds have been introduced — the latest one on June 30th, 2012.

> **86,164 seconds = 23 hours, 56 minutes and 4 seconds**
> **The duration of a star-based day on Earth (sidereal day)**
>
> **86,400 seconds = 24 hours**
> **The duration of sun-based day on Earth (synodic day)**
>
> **88,643 seconds = 24 hours, 37 minutes and 23 seconds**
> **The duration of a sidereal day on Mars**
>
> **88,775 seconds = 24 hours, 39 minutes and 35 seconds**
> **The duration of a synodic day on Mars, a 'sol'**

A leap day is something slightly different. To understand this, we need to take into account two effects. First, based on our calendar, we count 365 days in a year marking the time it takes for the Earth to circle the Sun once. In reality — as viewed from a faraway star — this actually takes 365.256363 days, or 365 days and 6.152712 hours (six hours and 9.16 minutes). This is called a sidereal year.

Secondly, the Earth's axis makes a precession movement, which resembles the whirling motion of a spinning top, albeit one that turns extremely slowly. Although the precession movement takes about 26,000 years to complete (see Chapter 14), it further affects the length of the seasons because they are defined by the position of the Earth's axis. A tropical year, or solar year, is defined by the length of four seasons; it is the time that the Sun takes to return to the same position in the cycle of seasons as seen from the Earth, namely 365.2421875 days, or 365 days and 5.8125 hours (5 hours and 48.75 minutes).

Therefore, our calendar year is a bit too short. To adjust it, a leap day was introduced every 4 years, on the February 29th. However, because the difference is not quite 6 hours but a little less, this would be a slight overcompensation. To correct that, it was decided that years devisable by 100, but not by 400, are not to be treated as leap years.

24 hours
One of the rhythms of our biological clock

In biology, we use the term 'circadian rhythm' to refer to daily patterns: the phrase stems from the words circa, meaning 'approximately', and dies, meaning 'day'. There are other biological rhythms that differ in duration. The menstrual cycle of women for example is a little less than a month (we come back to this in Chapter 8).

The biological clock is innate. It is driven by a small group of nerve cells in our brain, inside the hypothalamus. This group of nerves ensures that vital bodily functions occur within a regular time frame — such as blood pressure, heart beat, sleep cycles and body temperatures.

However, the regulation of our internal body clock can be influenced by external factors. One of those is the alternation between day and night,

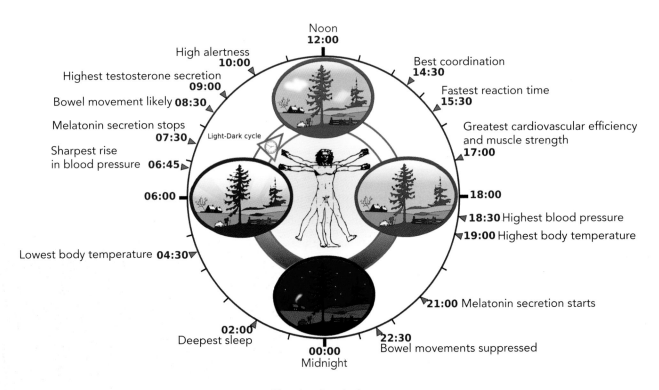

Noon
12:00

High alertness
10:00

Best coordination
14:30

Highest testosterone secretion
09:00

Fastest reaction time
15:30

Bowel movement likely **08:30**

Melatonin secretion stops
07:30

Light-Dark cycle

Greatest cardiovascular efficiency
and muscle strength
17:00

Sharpest rise
in blood pressure **06:45**

06:00

18:00

18:30 Highest blood pressure

19:00 Highest body temperature

Lowest body temperature **04:30**

21:00 Melatonin secretion starts

02:00
Deepest sleep

22:30
Bowel movements suppressed

00:00
Midnight

The circadian rhythm

light and dark. This is why we get confused when we travel long distances to eastern or western destinations, where the differences in time may be significant.

Our body temperature also displays a circadian rhythm. As we can see above, our temperature is highest at around 7 pm and lowest around 4.30 am. Another example of a bodily function with a circadian rhythm is the secretion of certain hormones.

86,400 seconds = 24 hours

Lifetime of a mayfly, also known in other languages as the 'one-day fly'

In some countries, the epitome of transience is the mayfly, which is better known as the one-day fly. Strictly speaking, a mayfly is not a fly at all, but a different type of insect (see picture on the right). They live for such a short period — from a few hours to a few days — that they do not even have a mouth or digestive system. After all, they do not have time to waste on eating or the usual forms of courtship, as they focus on mating and laying eggs.

The larvae of mayflies live in water, where they can actually stay alive for up to a year. Once emerged from

A mayfly

the water as mature insects, the mating ritual starts. The male dies after mating and once the female lays her eggs, she dies as well. Suffice to say their love life is short, but effective.

93,960 seconds = 26.1 hour
The half-life of ²⁰⁰Tl

Thallium is a silvery-white metal. It has atomic number 81 and its stable isotopes are ²⁰³Tl and ²⁰⁵Tl. The isotope in between (²⁰⁴Tl) is not stable and has a half-life of 3.78 years. Thallium-200 has 119 neutrons, the spin of its nucleus is 2, and the half-life is 26.1 hours. This isotope decays into mercury.

The chemical symbol for mercury is Hg, as it stems from the Greek and Latin word hydrargyrum, a contraction between hydro (water or fluid) and argentum (silver). Thallium is extremely poisonous. It was once used as a pesticide against rats and noxious insects, but that is no longer allowed.

You may have noticed that isotopes with an uneven number of neutrons often decay faster than their even-numbered counterparts. It is true that nuclei with an even number of neutrons and the same number of protons are the most stable. This is because protons and neutrons both have spin, in either direction around their axis. According to the laws of quantum mechanics, identical particles with a parallel spin hinder each other, which is why neutrons and protons with opposite spin like to be paired. If there is an uneven number of either, then a solitary proton or neutron is left to its own devices on an erratic path that destabilizes the nucleus.

144,000 seconds = 40 hours
For many of us, the duration of a working week

247,708.8 seconds = 68.808 hours =
2.867 days
The orbital period of the star Algol

Algol is the most luminous star in the constellation of Perseus. Even through the lens of an extraordinarily strong telescope, Algol looks like a normal star. In fact, it is a duo of stars circling each other (strictly speaking, there is a third star, but this one is not relevant for our story). One of the two stars, Algol A, is much brighter than the other one, Algol B. Seen from the Earth — sometimes with the naked eye — eclipses happen when one moves in front of the other.

During their periodic eclipses, the combined brightness of the stars decreases slightly. The light intensity is weakest when the darker star, Algol B, eclipses Algol A. The period between the same eclipses (for example Algol B eclipsing Algol A) is exactly 2 days, 10 hours and 49 minutes. The blackening out of Algol A is observable with the naked eye.

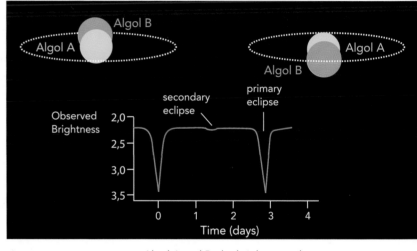

Algol A and B, the brightest and longest known of the so-called eclipsing variables, or 'Algol stars'.

Chapter 7

$10^6 = 1,000,000 = 1$ million

10^6 seconds = 11.57 days = 1.65 weeks

1 megasecond is the same as 1 million seconds. We are venturing into periods of between one and two weeks. The week consists of 7 days, which is 6.048 × 10^5 seconds. The concept of a week was introduced in the calendar by the Babylonians, who named each day after a deity. The creation of the world by God also took a week, of which the last day was a day of rest, the Sabbath. A year consists of 52 weeks, one or two of which might be spent on holiday.

The duration of the Apollo 11 mission in 1969 was 8 days, 3 hours and 22 minutes, including the 21.5 hours that the crew spent landing — and famously walking — on the Moon.

10 days
The universe after 10 days

Just 10 days after the Big Bang, the part of the universe that we are able to observe through our telescopes today had already expanded to 600 light-years. The temperature was still enormously high, about 12 million degrees Celsius. Because of the high temperatures, the mass density of 20 grams per cubic meter was still very small compared with the dominating energy of the omnipresent radiation. Matter — predominantly still atomic nuclei such as hydrogen and helium — comprised less than 1/1000th of the existing total mass-energy. These circumstances meant that gravity had very little grip on fluctuations, so the mass distribution of the universe must have been extraordinary homogenous.

11 days
The difference between a lunar year and a solar year

The duration of a solar month, defined as the period between consecutive full moons, is approximately 29.5 days. The oldest calendars are based on the movements of the Moon, and the word month (see also the next chapter) is derived from the word 'moon'.

A lunar year of 12 lunar months adds up to 354 days, 11 days shorter than the solar year, as defined in the previous chapter. The Islamic calendar continues to be based on the lunar cycle, which accounts for the fact that some holidays, like Ramadan, are always celebrated 11 days earlier than the previous year according to a solar calendar.

> **987,552 seconds = 0.99 × 10^6 seconds = 11.43 days**
> **The half-life of ^{71}Ge**

Germanium is a material that is frequently used in electronics to create semiconductors for transistor radios, among other things. Germanium has atomic number 32 and its stable isotopes are ^{70}Ge, ^{72}Ge, ^{73}Ge and ^{74}Ge.

It is remarkable that isotope ^{71}Ge, with 39 neutrons, is quite unstable with a half-life of 11.43 days. The product of ^{71}Ge's decay is stable gallium-71, with atomic number 31. The disintegration process does not follow the usual pattern, whereby a neutron is converted into a proton, electron and an anti-neutrino (beta decay), but via inverse beta decay. With inverse beta decay, an electron (e^-) close to the atomic nucleus is captured by its proton and converted into a neutron and a neutrino (ν_e).

This is presented schematically as follows:

$$^{71}Ge + e^- \rightarrow {}^{71}Ga + \nu_e$$

This reaction can also occur in reverse. By bombarding ^{71}Ga with neutrinos, we can produce ^{71}Ge, which in turn decays back into gallium:

$$^{71}Ga + \nu_e \rightarrow {}^{71}Ge + e^-$$

These reactions play an important role in the Sage Experiment, a Russian-American project to measure the number of neutrinos emitted by our Sun. The Earth is constantly being bombarded by neutrinos. These particles, which have no electric charge and only a very small — as yet unknown — mass, are created through nuclear reactions within the center of the Sun. They do not respond to electromagnetic force or gravity, but only to so-called weak interactions responsible for (inverse) beta decay.

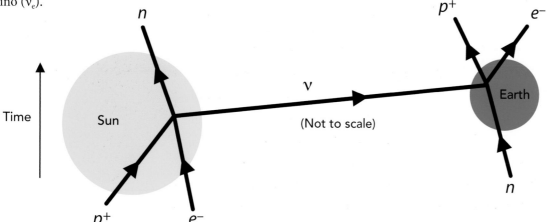

Inverse beta decay, triggered by neutrinos emitted by the Sun.

By exposing a large tank of gallium to the entering neutrinos from the Sun, germanium is produced, which then converts back into gallium. To protect the experiment against the influence of other cosmic particles, the tank is buried deep in the earth beneath the Caucasus Mountains. By measuring the production of germanium and its subsequent decay back into gallium, we gain an understanding of how many neutrinos are created by the Sun and reach the Earth.

This experiment has led to big, complicated questions regarding the nature of neutrinos, about which a lot of research continues to be conducted. For example, it has been discovered that there are various types of neutrinos, which can morph into one another.

13 days
Duration of the Cuban Missile Crisis during the Cold War

In October 1962, a nuclear war was looming between the United States of America and the then Soviet Union. The crisis started on October 14th when the United States deduced from spy-plane photos that the Russians were installing military bases in Cuba from which they could launch nuclear missiles.

On October 22nd the threat of nuclear war intensified, when President J. F. Kennedy imposed a blockade on the export of Soviet weapons to Cuba by setting up a blockade. He put the US Army on the highest state of alert.

Cartoon of Khrushchev and Kennedy's 'meeting of minds' over atomic weapons.

The tension reached its climax when Soviet ships came within the minimum possible distance of the blockade and a catastrophic war seemed inevitable. Behind the scenes, frantic negotiations were being held and after two days the countries reached an accord.

The Cuban Missile Crisis is considered the low point of the Cold War, and the incident was made into a movie in 2000 with the title *Thirteen Days*.

15.945 days
The orbital period of Titan around Saturn

Titan, the largest moon of Saturn, was discovered by Christiaan Huygens in 1655. The orbital period of Titan around Saturn is 15.945 days. In Greek mythology, the Titans formed a family of giants and Titan is indeed a gigantic moon. With its diameter of 5,150 kilometers — 40% of the diameter of the Earth — Titan is the second largest moon in our solar system, bigger even than Mercury and Pluto.

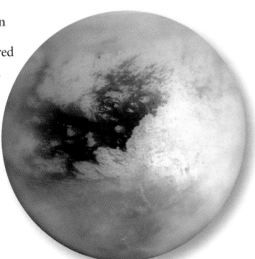

This picture of Titan was taken from a distance of approximately 115,000 kilometers, on 19 November 2007 by the Cassini spacecraft, in infrared light.

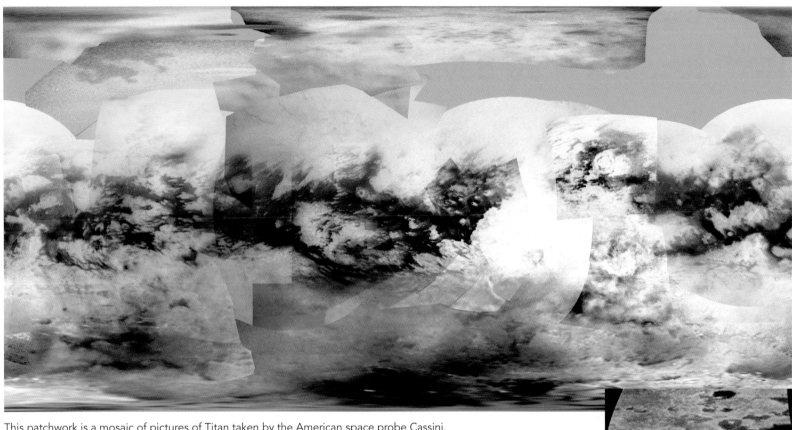

This patchwork is a mosaic of pictures of Titan taken by the American space probe Cassini. The dark spots are oceans and lakes of liquid methane, surrounded by land on which dunes have been identified. Inlay: radar picture of methane lakes on Titan.

With an average diameter of 5,268 kilometers, Ganymede is the largest moon in our solar system. It orbits Jupiter in 7.155 days.

On 14 January 2005, a spacecraft named after Huygens entered Titan's atmosphere and landed on the moon. The Cassini-Huygens probe was designed specifically for its mission to Saturn and its moons.

The space probe consisted of two parts, of which the smallest part, Huygens, could be detached to study Titan. The other part of the space probe, Cassini, was built to study Saturn itself and all its moons. Perhaps the most interesting fact about Titan is that it has a thick atmosphere. In fact, it is the only moon in our solar system with an atmosphere at all. This means that Titan has its own climate, clouds are formed and sometimes it even rains. But it is very cold on Titan, about 179 degrees below zero, so when it rains, it is liquid methane that falls from the sky, not water.

31

Chapter 8

$10^{6.41} = 2{,}592{,}000$
$10^{6.41}$ seconds = 30 days $\approx$ 1 month

In this chapter, we digress from the integer powers of 10. However, as so many elements of our daily existence are related in one way or another to the period of a month, we decided to dedicate a chapter to this time span anyway. In terms of powers, we are somewhere between the sixth and seventh power of 10. A month of 30 days consists of exactly 2,592,000 seconds = 2.592×10^6 seconds, thus almost 2.6 million seconds. We can also denote this as $10^{6.41}$ seconds.

A month is a twelfth of a year. As the word suggests, its original meaning was derived from the orbital period of the moon. Just like a 'day', there is a sidereal or star-month as well as a synodic month (discussed in more details in the main text). The first calendars were based on the phases of the Moon, but the problem arose that the Moon, as seen from the Sun, makes a little more than 12 rotations around the earth each year. Therefore, a new system was devised using the Sun as the basis for the calendar. As a consequence, the months had to last a little bit longer: 30 or 31 days — with the exception of February having 28 or 29 days.

Lunar calendars are still based on the synodic month. The Islamic festival of Ramadan lasts exactly one lunar calendar month (see also Chapter 7). Some Christian holidays are also based on the lunar calendar — Easter for example, falls on the first Sunday after the first full Moon in spring.

The origin of the various months and the definition of the number of days they contain lie in Roman times. Before 153 BC, a year started with the month of March, which is why February, the last month of the year in those times, became somewhat shorter.

The Roman dictator Julius Caesar introduced a new calendar, which was named after him. Once every four years, the month of February was to have 29 days rather than 28 days, in order for the seasons to start on the same calendar day every year.

In Caesar's honor, the Roman senate decided to rename the fifth month of its calendar from Quintilis (also sometimes spelt Quinctilis) to Julius. The four preceding months had already been named after important deities: Mars, Aphrodite*, Maia (the goddess of spring), and the queen of the Roman pantheon, Juno. When, 37 years later, Emperor Augustus managed to reestablish Rome as a flourishing city, and the Senate identified that many of his conquests had taken place in the sixth month (Sextilis), it was decided that it should be called Augustus. Subsequently, several attempts were made to name other months after Roman emperors as well: April was to be called Neroneus, after Emperor Nero, and May was to be named after Claudius. However, these attempts never caught on. In medieval times, the beginning of the year was chosen differently in different countries.

Caesar's calendar proved to be so accurate that it was used throughout Europe until October 4th, 1582. It was clear to 16th century scholars that the seasonal equinoxes were falling ten days 'too early'. An improved system, introduced by Pope Gregory XIII, was adopted by Catholic countries in Europe. Caesar's calendar had given the year 365.25 days, whereas the actual duration for the Earth to orbit the Sun is almost 11 minutes shorter, a difference that amounts to three days in 400 years (see also Chapter 6). Therefore, Gregory XIII posited that years divisible by 100 should not be leap years, unless they are multiples of 400, such as the year 2000, which is a leap year again.

Other countries adopted the Gregorian system at different times, so that different numbers of days (from 10 to 13) had to be removed from their calendars. In England and its colonies, the new system was adopted in 1752. There, September 2nd, 1752 was followed by September 14th, 1752. After that, several other reforms were attempted, in order to rationalize the system. During the French Revolution, for instance, the year was divided into 12 months of 30 days each, and the month was divided into three 'decades' rather than weeks. A few days of festivities were added to give the year 365 or 366 days. This calendar lasted from 1793 to 1805, when Napoleon reinstituted the Gregorian calendar.

2.194556 × 10⁶ seconds = 25.4 days
Rotational period of our Sun

The Sun is the energy source of our solar system, including our planet Earth. Eight planets circle the Sun, each with their own orbital and rotational period. The Sun also turns around on its axis, with a rotational period of 25.4 days. Its rotational period was determined by observing the surface of the Sun, which shows several 'sunspots', which are darker because their temperature is lower (see also Chapter 11). Because of the Sun's rotation,

---

*Whether the name of the month April indeed originates from the goddess Aphrodite is debated; researchers argue that the Romans would have used the name Venus, and the origin of the name April remains unclear.

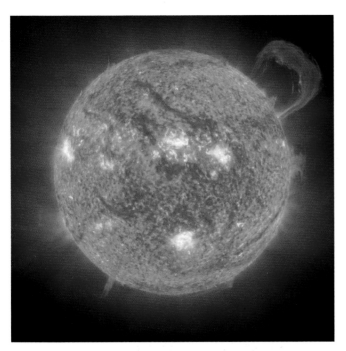

The Sun

lations have shown that the Sun is approximately 4.5 billion years old. This is also the approximate age of the other planets in our solar system. The Sun is an enormous fireball, consisting mostly of hydrogen that converts to helium through nuclear reactions. This creates energy in the form of heat radiation, which provides the Earth with enough heat to make it habitable.

$2.360595 \times 10^6 = 27.3217$ days
Duration of a sidereal or stellar month

This is the time it takes for the Moon to orbit the Earth and return to a fixed point that is measured against the stars (see figure below).

we can identify the dark spots passing by us and that's all we need to deduce its rotational period.

Not all parts of the Sun, however, turn at the same speed; at the equator it rotates in 25 days, while the rotational period around the poles is closer to 36 days. Differing rotational periods for various parts of a sphere is, of course, only possible with a gasiform or fluid body — while in a rotating solid object, all its parts must have the same rotational period.

The Sun is only one of many, many stars in our galaxy, called the Milky Way — somewhere between 200 and 400 billion. Of all these stars, the Sun is the one closest to us, only eight light minutes away, or 150 million kilometers. Calcu-

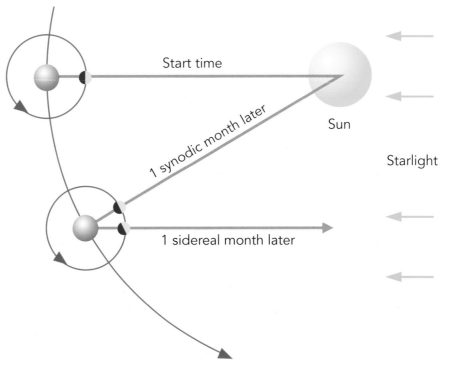

The difference between a sidereal and synodic month.

$2.551444 \times 10^6 = 29.5306$

Duration of a synodic or solar month

This is the time it takes for the Moon to complete its full cycle of phases as viewed from the Earth: an average of 29.5306 days between new moons = 29 days, 12 hours, 44 minutes and 2.8 seconds. As the Earth is also moving along its orbit, the Moon needs more time to regain its position in a straight line with the Earth and the Sun: this requires two days more than a sidereal month.

25 to 35 days

The duration of a menstrual cycle

A number of biological clocks regulate the human body, two of which have already been discussed: the heartbeat in the first chapter and the circadian rhythm in Chapter 6. In the female body, another biological clock exists: the menstrual cycle, which typically lasts around four weeks. In the initial phase, the ovum (or egg) develops in the ovaries under the influence of hormones as directed by the brain. This takes about two weeks. The ovum is situated in a vesicle filled with fluid: a follicle. The follicle releases the ovum, which pushes out of the ovaries after one or two days into the oviducts, more commonly known as fallopian tubes. Through muscle contractions and the efforts of tiny hair-like cilia inside the oviducts, the ovum is moved to the uterus, where it nestles into the endometrial layer of blood in anticipation of possible insemination. If no insemination occurs, the ovum leaves the uterus, along with the bloody womb lining. This bleeding continues for about five days, and is referred to as the menstrual period.

It is probably not a coincidence that the menstrual cycle coincides with the rhythm of the moon, but though the subject has been hotly debated and rigorously tested, no clear connection is yet proven. One might suspect that, in prehistoric times, the moon must have been very important socially, because of its lightening up of the night sky.

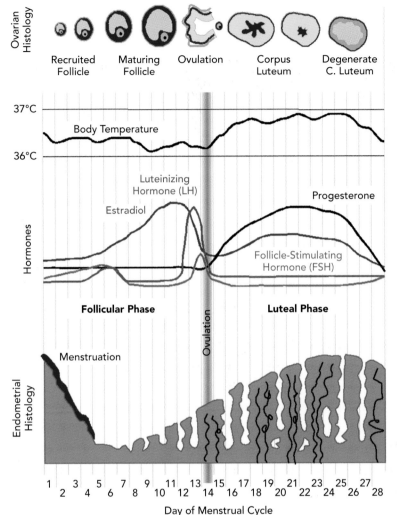

(Average values. Durations and values may differ between different females or different cycles.)

The menstrual cycle. On the vertical axis at the top, the levels of the hypophyseal hormones, FSH (follicle-stimulating hormone) and LH (luteinizing hormone) are pictured.

27.7025 days = 27 days, 16 hours and 52 minutes
The half-life of ^{51}Cr

Chromium, with the atomic number 24, is a metal that exists quite commonly in the Earth's crust. Mostly occurring in chromium compounds, we can also find deposits of the metal itself. Chromium-51 is radioactive because it transmutes into ^{51}V through electron capture and the emission of a photon (^{51}V is the only stable isotope of vanadium). ^{51}Cr is used in medical research to label red blood cells, through which their number and lifespan can be identified.

One month and six days
The duration of the first crossing of the Atlantic Ocean, by Christopher Columbus

His epic voyage commenced on September 6th, 1492 from the Canary Islands, with a crew of 90 men on three ships: the Santa Maria, the Pinta and the Niña. One month and six days later, on October 12th, 1492, one of the sailors yelled 'land ahoy!'. The land proved to be one of the islands of the Bahamas, just east of the American state now called Florida. Columbus, believing he had reached the East Indies — as the entire Southeast Asian area was called at the time — named its inhabitants 'indios', Spanish for 'Indians'.

Scale model of the Santa Maria, one of three ships under Columbus's command.
The other two were the Pinta and the Niña.

Chapter 9

10^7 = 10 million
10^7 seconds = 115.74 days = 3.86 months

This is the timescale of the seasons. Ten million seconds is almost four months, and one year is 3.1536×10^7 seconds. The period of three months is also referred to as a quarter — many companies report their financial results 'quarterly'. In higher education in continental Europe, a course is often taught for one-quarter of the year.

The seasons are based on the motion of the Earth around the Sun. During the four seasons we observe different environments on the northern and southern hemispheres. Differences in climate are determined to a large extent by the inclination of the Earth's axis with respect to the surface of the orbital plane around the Sun. In summer time, the northern hemisphere is exposed to more and longer episodes of sunlight and, as such, is relatively warm. In the winter, the opposite is true, and the southern hemisphere enjoys the Sun's energy. On (or near) June 21st, the Sun is positioned perpendicular to the Tropic of Cancer (one of the five major circles of latitude on maps of the Earth), the point in time that marks the beginning of summer for the northern hemisphere. Around December 22nd, the Sun is perpendicular to the Tropic of Capricorn, the start of our winter.

Both spring and fall commence when the day and night are equally long, the so-called vernal and

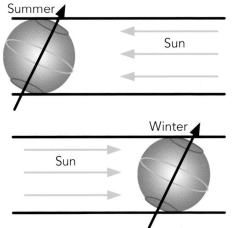

The northern hemisphere receives more energy from the Sun in the summer than the southern hemisphere; it will be summer in the northern hemisphere and winter in the southern hemisphere.

autumnal equinoxes. This is when the Earth's axis does not point towards the Sun or away from it, but is perpendicular to an imaginary line running between the Earth and the Sun. This is illustrated on the next page. At this point in time, the Sun is perpendicular to the Earth's equator. Because of small variations in the movement of the Earth's axis, the inclination varies as well and, if we were to be precise in regards to the

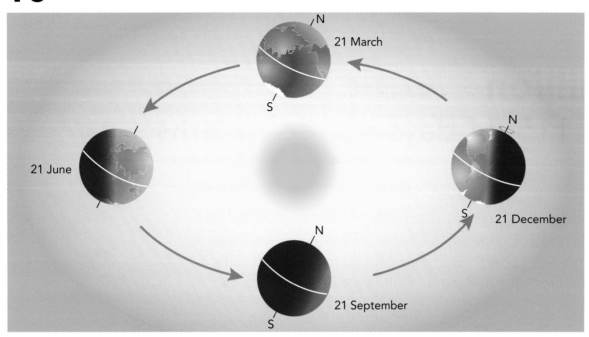

N

21 March

S

21 June

N

21 December

S

N

21 September

S

The inclination of the Earth's axis determines seasonal variations in its annual orbit around the Sun.

Earth's motion, precession movements need to be taken into account too (elaborated upon in Chapter 14). This is why the vernal equinox, the beginning of spring, does not always fall on the same day, but sometimes starts on March 20th or 21st.

It should also be mentioned that, during the equinoxes, the days actually last somewhat longer than the nights, and at both poles, there is no night at all. This is because of two effects: the Sun is not a point but a circle so that parts of it stay visible a bit longer, and the Sun rays get bent around the Earth due to the atmosphere: the Sun appears to occupy a slightly higher position in the sky than where it actually is.

0.76 × 10⁷ seconds = 88 days
The orbital period of the planet Mercury around the Sun

Mercury is the closest planet to the Sun. During the day it can be very hot on Mercury, up to 427°C. At night the planet cools down considerably, to −183°C. The distance between Mercury and the Sun varies between 50 and 70 million kilometers, which is almost three times closer than the Earth to the Sun. Mercury's orbit is rather eccentric and strongly elliptical, which explains the variance in distances. Looking up into the sky from Mercury, the Sun would move rather strangely along the horizon.

A 'year' on Mercury lasts 87.969 days — very short in comparison to our year on Earth, but the rotational period of Mercury is very long by contrast: 58 days, 15 hours and 30 minutes. A quick calculation shows that the planet's orbital period around the Sun is 1.5 times its rotational period. This means that when Mercury's position is closest to the Sun, the same side of the planet is exposed for a very long time. After one orbit, Mercury turns around its axis 1.5 times so that the exact opposite side of the planet is turned towards the Sun. This is a phenomenon that is well known in astronomy: Mercury's rotation around its axis is linked to its orbital motion. (The rotation of our Moon is also linked to its orbital motion, which is why the Moon always shows the same face to us.) Linking happens when a body deviates from a perfect sphere. Mercury, and our Moon, are not perfect spheres. The Sun holds Mercury in its grip, but because Mercury's orbit is so elongated, the planet's orbital distance and velocity vary, so that it tumbles around while orbiting.

Because the rotational period is only a little shorter than its orbit around the Sun, Mercury's solar days are quite long: 176 Earth days or exactly two Mercury years. Mercury, with a diameter of 4,880 km, is the smallest planet in our solar system. Pluto is even smaller, but was demoted to a 'dwarf planet' in 2006 and thus doesn't count.

The exact location on Mercury's orbital path where it is closest to the Sun, called the perihelion, shifts over time. This precession baffled scientists for a long time, because the measured size of the shift could not be explained by the planetary laws of Kepler or Newton's laws of motion. Only in the 20th century was Albert Einstein able to explain the deviation with his new theory on gravity — the general relativity theory. His revolutionary theory fundamentally changed our insight into the concepts of time and space.

106.64 days
The half-life of ⁸⁸Y

At atomic number 39 on the periodic table is the chemical element yttrium. It is a silvery-white transition metal. Natural occurrence of the element on Earth is rare. ⁸⁹Y is stable with 50 neutrons, but with one neutron less, ⁸⁸Y decays through electron capture (see also Chapter 7) into strontium-88 with atomic number 38. Through this process, a proton is turned into a neutron — the inverse process of beta decay, as discussed in earlier chapters.

The stable isotope of yttrium plays a prominent part in manufacturing superconducting materials. In superconductors, resistance in electric currents disappears completely. A much-explored and used superconductor is based on the

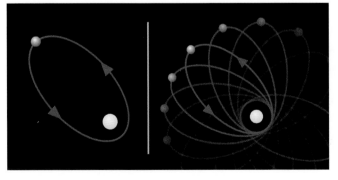

Left: the perihelion of a planet in an elliptical orbit around the Sun. Mercury completes its elliptical path in 88 days. On the right, the shift in location of the perihelion is shown, but is highly exaggerated.

chemical compound yttrium-barium-copper-oxide (more accurately presented as $YBa_2Cu_3O_7$), which becomes superconducting at a temperature of 96K (or −177°C). The explanation as to why superconductivity occurs at these relatively high temperatures remains a mystery — scientists are working hard to find a fitting theory to explain this phenomenon.

119.779 days
The half-life of ^{75}Se

Selenium has atomic number 34 and its isotope ^{75}Se has 41 neutrons in its atomic nucleus. ^{75}Se decays into the element arsenic, which is ^{75}As with atomic number 33 — in other words, one proton less, one neutron more. Its decay occurs through electron capture, because there is not enough energy available to emit a positron. The other electrons sink towards shells with a lower energy level, creating radiation in the process. In the case of ^{75}Se, the radiation is highly energetic (and thus radioactive), a phenomenon also referred to as gamma radiation.

^{75}Se is used for applications in radiography, for example, for a technique used to identify minuscule cracks in material that cannot be observed by the naked eye. These faults are made visible because gamma rays have a larger penetration depth than radiation with a lower energy — by measuring the absorption of the radiation, variations can be identified that indicate cracks deep inside the material.

A domestic sow suckling her piglet.

115 days

The average gestation period of pigs

Gestation periods for mammals vary quite widely. For humans this period is approximately nine months or 275 days. Longer gestation periods are found in horses (336 days) or elephants (more than 600 days). In the category 'quick breeds' we find rabbits (33 days), cats (62 days) and dogs (65 days). Pigs are somewhere in the middle with a gestation period of around 10^7 seconds.

10^8 = 100 million
10^8 seconds = 3.17 years

Within three years, a child grows from a newborn to a toddler. After about a year the baby has already learnt how to laugh, walk and he may even babble a few well-chosen words. After that he masters eating with a spoon and perhaps works out how to pile up a few building blocks without them toppling over. At around the age of three, many kids are potty trained and go to kindergarten.

Three years is also the typical duration of a Bachelor's degree program at university. It took us about three years to write this book. Also in three years, the five ships in the fleet of Ferdinand Magellan circumnavigated the globe for the first time ever (1519–1522). The Portuguese explorer, who perished during this voyage, gave his name to a vital waterway through Chile that links the Atlantic and Pacific oceans, two lunar craters and one on Mars, and a Patagonian penguin.

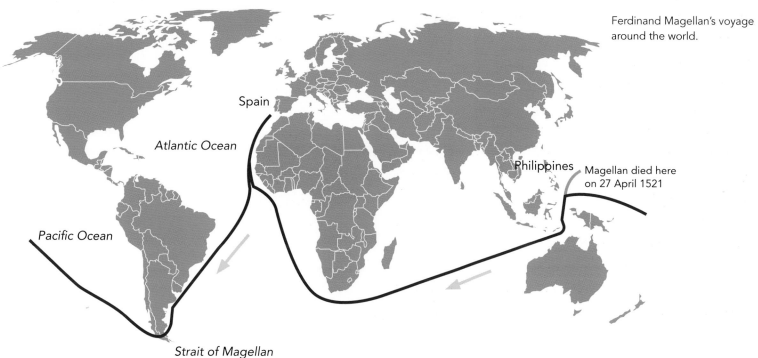

Ferdinand Magellan's voyage around the world.

Spain

Atlantic Ocean

Pacific Ocean

Philippines

Magellan died here on 27 April 1521

Strait of Magellan

3.16 years
The orbit of asteroid 13092 Schrödinger

Between the planets Mars and Jupiter lies a large asteroid belt, which contains an estimated one to two million rocks with a diameter of at least one kilometer. The largest of these asteroids is the dwarf planet Ceres, discovered in 1801, with a diameter of about 1,000 kilometers.

While there is a huge number of asteroids here, their total combined weight is estimated to be only 4% of the Moon's mass. In other words, they are relatively small bits of debris that orbit the Sun in quite stable trajectories.

They date to the very origin of the solar system, when planets were formed as a result of clotting of planetesimals. However, because of Jupiter's strong gravitational force field such clotting did not occur here, and so they continue to orbit the Sun as individual pieces of debris.

The picture on the right appears to suggest that the asteroids are quite close to each other. It only seems so from a distance and spacecraft are able to navigate through the belt without any problems. That said, relatively large asteroids do crash into each other at times. A number of asteroids exist closer to Earth — the chance that one of them will actually collide with us is small, but is nonetheless a good enough reason for us to monitor the belt closely.

The orbital times of the asteroids in the asteroid belt vary from 2.5 to 8 years. In comparison, the orbital time of Mars is 1.88 and that of Jupiter

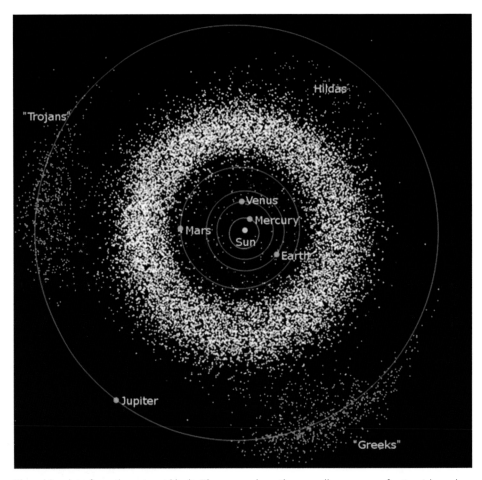

The white dots form the asteroid belt. There are also other, smaller groups of asteroids such as the Trojans and Greeks, whose positions are stagnant in relation to Jupiter.

Asteroid Ida (completes an orbit in 4.84 years and has a rotational period of 4.63 hours) even has a small moon, Dactyl. The picture was taken by the American spacecraft Galileo, on its way to Jupiter.

11.86 years. Many of the asteroids are named after well-known scientists, musicians and artists, and other current or historical celebrities. The asteroid 2001 Einstein is a fast one, with an orbital time of 2.69 years. While 1818 Brahms takes 3.18 years to complete its orbit, 4330 Vivaldi needs 3.36 years and 1034 Mozartia, 3.47 years. The Dutch Nobel laureates for physics in 1999, 't Hooft and Veltman, each received an asteroid named after them, with an orbital period of 3.29 and 3.23 years, respectively. Meanwhile, 13149 Heisenberg takes 5.54 years to orbit the Sun, and 4317 Garibaldi, which is much further away from the star, has an orbital period of 7.96 years. Closest to a power of 10 is the orbital time of the asteroid Schrödinger, 3.16 years or about 10^8 seconds. Schrödinger was an Austrian physicist, well known as one of the founders of quantum mechanics.

3.3 years
The half-life of ^{101}Rh

Rhodium is a silvery-white transition metal. In the periodic table of chemical elements we find it under atomic number 45, meaning it has 45 protons in its nucleus. It has good light-reflective qualities and is therefore commonly used to give jewelry a beautiful shine. As it does not occur in large quantities in natural form, it is quite expensive. It is also used in the automotive industry, as a catalyst for cleaning exhaust fumes. In its stable form, ^{103}Rh, the nucleus contains 58 neutrons. All other isotopes — and there are about 20 — are

unstable. ^{101}Rh is the longest living with a half-life of 3.3 years. It decays through electron capture into ruthenium-101.

Four years
An Olympiad, the period between two Olympic Games

The idea to hold recurring sporting games once every four years was launched in 776 BC in the Greek sanctuary, Olympia. When the Roman emperor Theodosius I prohibited the games in 394 AD, it is believed that some athletes defied the ban and continued to compete, but after a while the games died out. Only in 1896 was the tradition restored, with Athens hosting the first modern Olympics.

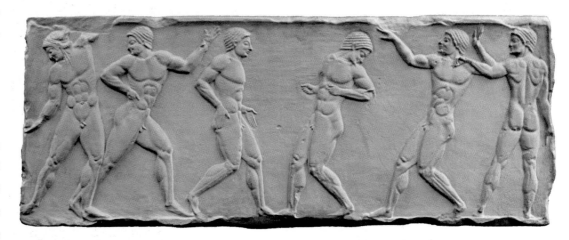

4.22 years
The time a light ray needs to travel from the Sun to the nearest star

Our Sun is one of 200–400 billion stars in the Milky Way. The exact number is anyone's guess. We do know, however, the total mass of the Milky

Way, which has been measured quite precisely at 700 billion solar masses. This number can be derived from velocity measurements of stars at the outer edge of the galaxy, knowing that the orbital velocity is determined by the total mass inside the orbital radius. But we do not know what percentage of that mass consists of stars and how much is 'dark matter', or which stars are so small and light that they are (too) difficult to detect.

Proxima Centauri (*proxima* is Latin for 'near') exists in the constellation of Centaurus. This is the star that is nearest to our Sun, at a distance of 4.22 light-years. By comparison, the distance between the Sun and the Earth is only eight light-minutes (see Chapter 4).

The images below compare the distances in kilometers to the Sun. We see that there are two other stars close to Proxima Centauri: Alpha and Beta Centauri. The latter two form a so-called double star, a system of two stars circling each other.

The distance between Alpha and Beta Centauri is a little greater than the distance between our Sun and the planet Uranus, in other words about half the size of our solar system. Alpha Centauri is the brightest star of the trio. Proxima Centauri is closer to us, but no more visible. The distance between Alpha and Proxima is about 0.15 light-years. There are uncertainties as to whether Proxima actually orbits Alpha and Beta Centauri. Astronomers calculate that its orbit must be enormous, about a million years. If this is correct, then there will be a moment in time — in about 100,000 years — when Alpha Centauri will be the closest to us. Alpha Centauri will then be the 'proxima'.

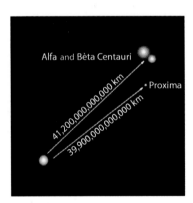

Binary system Alpha and Beta Centauri, with Proxima on the right-hand side, the star closest to the Sun. Expressed in kilometers, the distances to the Sun are enormous. Space travel to stars is impossible for humans because our current spacecraft would need tens of thousands of years to get there. Even light rays from the Sun, which travel with a velocity of 300,000 kilometers per second, take 4.22 years to reach the nearest star!

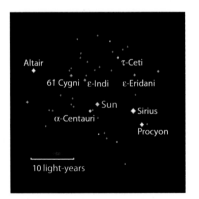

A few of the closest stars in our Milky Way. The typical distance between two of these is a few light-years.

Chapter 11

10^9 = 1 billion

10^9 seconds = 31.7 years

The period of 10^9 seconds is also called a gigasecond. In one gigasecond, more than 30 years will pass. 'Giga' comes from the Greek *gigas*, which means giant or gigantic. In IT, the term 'giga' is often used in the word 'gigabyte'.

The name 'gigabyte' is not strictly accurate. A gigabyte is really 2^{30} = 1,073,741,824 bytes, which is not exactly the same as 10^9. The term gigabyte is therefore not used in the correct manner, which is why the alternative term of *gibibyte* was introduced to indicate a storage capacity of 2^{30} bytes.

The computer, by the way, has changed significantly in one gigasecond. The first Apple saw the light of day in 1976 (see picture on the right), containing a RAM memory of four kilobytes — or rather four kibibytes = 4,096 bytes. As little as one gigasecond later, most personal computers have a memory of at least 64 gibibytes — and quite a bit more by the time you are reading this book.

As we reach 9 powers of 10, we encounter potential communication problems: depending on where you come from, you might consider 10^9 to be a billion, a thousand million or even a milliard. In a separate section after the introduction of this book, we have summarized the long and short scale nomenclature; see also the beginning of Chapter 17.

The evolution of the Apple computer over 30 years. The Apple 1 (above) was sold as a 'do-it-yourself' assembly kit, without the homemade wooden casing (below).

29.45 years
The orbital time of the planet Saturn

Saturn, with its eye-catching and attractive rings, is the largest planet in our solar system after Jupiter. It can be seen with the naked eye, and beautiful pictures of the planet were taken recently by the spacecraft Cassini. A full circle around the Sun takes Saturn almost 30 years. The planet has a gas and liquid-like composition and turns around its axis in about ten hours. The planet's poles are a bit flattened. The origins of its rings continue to be a subject of discussion, but it is suspected that they are the result of a moon that broke up when it came too close to Saturn. The many pieces of debris show varying rotational periods: particles closer to Saturn turn more quickly around the planet than those further away. Detailed observations have shown that it is actually a system of seven rings with complex inner structures due to the effects of gravity from other moons in the vicinity (see figure above). Saturn has quite a large number of moons: 62 have been counted so far. Titan is the largest, comprising about 90% of the mass of all the moons orbiting Saturn, including its rings. Titan is quite a bit larger than our own moon and even has an atmosphere. More about Titan can be found in Chapter 7.

The planet Saturn, the impressive gas giant, with its rings and moons.

30 years

The Thirty Years' War started in 1618 and ended in 1648

Initially, war broke out between Catholics and Protestants in the Holy Roman Empire, predominantly involving German states. The cause of the war was the so-called (Second) Defenestration of Prague, when two Catholic advisors of Emperor Matthias (Archduke of Austria and King of the Holy Roman Empire) were thrown out of the window of a castle in Prague by Protestant noblemen. (The word 'defenestration' comes from the French '*de fenêtre*', which means 'out of the window'.) The conflict spread to other states of Europe, politicizing the war. The Thirty Years' War can be seen as a concatenation of different conflicts between various states at different times.

In 1648, the war ended with treaties between the Holy Roman Empire and several European states, commonly referred to as the Peace of Westphalia. Essentially it meant the end of the Holy Roman Empire's strength and dominance, and recognition of a number of Protestant and Calvinistic states — it was the first time, for example, that the Republic of the Seven United Netherlands was officially recognized.

30.07 years
The half-life of ^{137}Cs

Cesium is a silvery-gold metal, with atomic number 55. ^{133}Cs, with 55 protons and 78 neutrons, is stable and is commonly used in atomic clocks, which we discussed at length in Chapter 1 and in Chapter 38. With four extra neutrons in the nucleus we find ^{137}Cs, an extremely toxic and radioactive isotope. Because of its relatively long half-life of 30.07 years, ^{137}Cs is a significant hazard to the environment when exposed. Even up to the present day it continues to be a dangerous source of radioactive radiation in the area of the old nuclear reactor of Chernobyl (Ukraine), which exploded in 1986 (see figure below). Cesium-137 is produced in nuclear fission, and decays into barium-137. ^{137}Ba has a half-life of 2.55 minutes, but emits extremely hazardous radiation during its decay.

Nuclear reactor of Chernobyl

Pope Pius IX (1792–1878)

32 years
The tenure of Pope Pius IX

The length of a person's working life averages between 30 and 40 years, but this is of course not true for all professions. Pope Pius IX was the longest-reigning pope in the history of the Catholic Church (Saint Peter, the apostle considered by many to have been the first pope, not included). The pontificate of Pope Pius IX began in 1846 and lasted until his death in 1878 aged 86.

He is famous for convening the first Vatican Council (1868–1869), which decreed the dogma of the pope's infallibility. Before that, he also introduced the dogma of the Immaculate Conception of Maria, the mother of Jesus. This meant that her soul, and thus also Jesus's soul, is never fouled by sin, not even by the 'original sin' committed by Adam and Eve and inherited by their progeny. However, Immaculate Conception, often celebrated on 8 December, is not recognized by Protestant or Orthodox religions.

33.227 years
The orbital period of the comet Tempel-Tuttle

Comets are some of the most fascinating objects in the sky. They consist of a nucleus and a tail. The nucleus is like a dirty snowball a few kilometers in size. The tail is made when the comet nears the Sun, which causes the ice to melt and evaporate. Dust and vapor escape from the center and are blown away by the Sun's winds. Comets' orbital periods range from a few years to more than 100,000 years. To date, a few thousand comets have been identified, but it is suspected that many millions exist. Most of them are invisible because they pass by far beyond Pluto's orbit.

In 1865–1866, Ernst Tempel and Horace Tuttle identified a comet with a relatively short orbital time of 33 years. The last passage of the Tempel-Tuttle comet was in 1998, but the comet was not visible with the naked eye. It belongs to the short-period comets (meaning with an orbital period of less than 200 years) from the so-called Kuiper belt. The famous Halley's Comet, with an orbital period of 76 years, passes the Earth closer than Tempel-Tuttle, which means it is brighter. As opposed to Tempel-Tuttle, Halley is visible with the naked eye (just like comet West, seen on the right).

Comet Tempel-Tuttle is considered to be the source of the Leonid meteor shower, a group of meteorites named after the constellation Leo, from which they appear to originate. The comet emits meteor particles after it has been near the Sun. The Earth passes through this debris, causing a cascade of particles to bombard the Earth's atmosphere — sometimes with a combined weight of more than ten tons of matter. Their combustion in the atmosphere looks like a shower of falling stars. The spectacle occurs every year in November, when observers can see hundreds of shooting stars in an hour.

Comet West, discovered by Richard West of the European Southern Observatory, in August 1975.

33.7 light years
The distance to the star Pollux

Castor and Pollux are two figures from Greek mythology. Supposedly the twin sons of the god Zeus, they formed an inseparable duo. In the sky, we find them in the constellation Gemini, which means 'twins'. In this constellation, Castor and Pollux are two stars that form the heads of a couple holding hands, as can be seen in the figure below. They are the two most luminous stars in the constellation — with Pollux slightly brighter than Castor.

Light from the star Pollux takes 33.7 years to reach the Earth. The star is almost twice as heavy as our Sun and has a diameter of about eight times the Sun's size. Since 2006, we are aware that

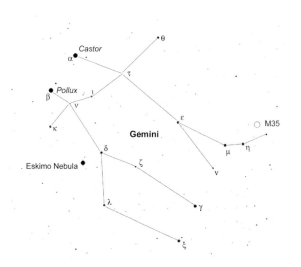

Castor and Pollux, the two brightest stars in the constellation Gemini.

an enormous planet orbits Pollux, aptly named Pollux b. This exoplanet (so named because it is a planet outside our solar system) is at least twice the size of Jupiter, and orbits at a distance of 1.5 AU (Astronomical Unit, the average distance between the Earth and the Sun) from its parent star. Because Pollux currently has a brightness (radiation power) of about 32 times that of the Sun, it is unlikely that any life is possible on Pollux b. It is not impossible, however, that there may have been life on the planet in an earlier phase of the parent star.

Every year in the month of November meteorites or falling stars, which appear to originate from the constellation Leo (Lion), are visible in the sky — the Leonids.

Chapter 12

10^{10} = 10 billion
10^{10} seconds = 317 years

Three hundred years ago, the end of the 'Golden Century' was nearing for Holland — the time of great statesmen and Dutch colonialism. The Netherlands reigned over what is now called Indonesia for more than three centuries, from 1602 to 1949. Within its own borders, it was also the time of great Dutch scientists, writers and painters. The famous philosopher Spinoza died in 1677, scientist Christiaan Huygens in 1695, and the painters Vermeer and Rembrandt in 1675 and 1669 respectively. Joost van den Vondel wrote his renowned play *Lucifer* in 1654, and the philosopher and scientist René Descartes resided in the Netherlands from 1628 to 1649.

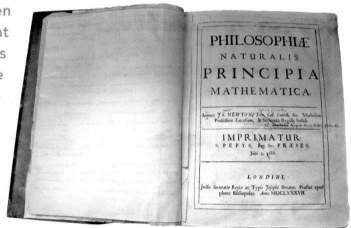

The *Principia* by Newton

The English physicist Isaac Newton, considered by many to be the greatest scientist of all time, also lived contiguously with these masters of the 17th century. He wrote his best known work, the *Principia*, in 1687, thus more than 10^{10} seconds ago. His masterpiece includes his theory on gravity and describes the laws of classical mechanics. The works of these leading historical figures continue to be of great significance for society today.

0.78×10^{10} seconds = 248.54 years
The orbital period of Pluto

Pluto is no longer considered to be a planet, having been demoted to a mere dwarf. In recent years, many such dwarf planets have been identified far away in our solar system. Pluto is between 4.4 and 7.4 billion kilometers from the Sun. This is why the average temperature on Pluto is very low, about 229°C below zero. The dwarf planet was discovered in 1930, but it took another 25 years

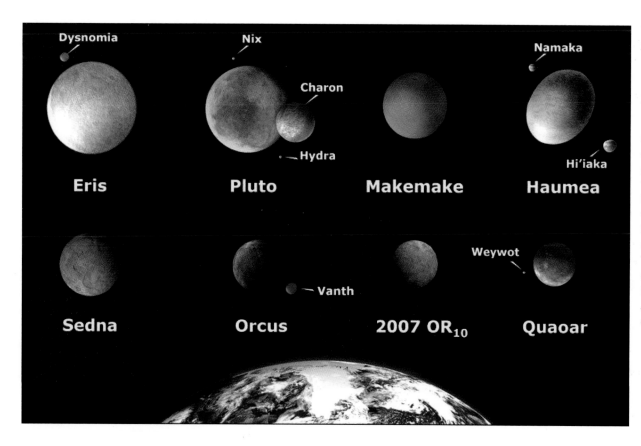

The eight largest known trans-Neptunian dwarf planets and their moons, at scale. Pluto and its moon Charon are closest to the Earth. The pictures are, in fact, impressions, because it has not been possible to observe any of these dwarf planets in great detail. In July 2015 or thereabouts the spacecraft New Horizons (see also Chapter 5) will reach Pluto. It will then be able to send much better images to us. There is also a dwarf planet closer to home, Ceres. Ceres is located between Mars and Jupiter, and is the largest object in the asteroid belt.

to determine its orbital period. Charon, Pluto's first moon, was discovered in 1978 and in 2005 the Hubble telescope discovered two more moons, Hydra and Nix. Today, the number of its moons has risen to five.

The largest dwarf planet, Eris, with an orbital period of 557 years, was found in 2005. Eris is larger and heavier than Pluto, but is about three times further away from the Earth. If Pluto ever deserved to be called a planet, then Eris should have been one too. Another dwarf planet is

Makemake, with an orbital period of 310 years. It is the third largest dwarf planet in our solar system after Eris and Pluto. All of them belong to the Kuiper Belt which, as with the asteroid belt (see Chapter 10), contains many thousands of objects.

The dwarf planet Sedna, whose origin is yet unknown, is the furthest away from the Sun. It has an orbital period of about 12,000 years. It is extremely cold on Sedna, at minus 240°C . Sometimes the temperature falls to only 15 degrees above absolute zero (−273°C).

200–500 years

The longest living animals

We know that many land tortoises are at least 150 years old. Adwaita is the name of a giant tortoise weighing 250 kilograms, who died in 2006. According to folklore, Adwaita was born in 1750, meaning it died at the age of 256. Reports suggest the tortoise was caught by British sailors in the Seychelles and grew up as a pet animal of General Robert Clive of the East India Company. It was then the longest surviving animal in a zoo in Calcutta. Adwaita's real age continued to be the subject of discussion, until research with the aid of radiocarbon dating (see Chapter 13) confirmed Adwaita's impressive longevity.

Tortoises are not the only animals that can live for centuries. Whales from Greenland and certain types of Japanese koi can reach the grand old age of 200, or even more. But a species of hard-shelled clam (*Arctica islandica*), also known as ocean quahog, takes the crown when it comes to age: in 2012, scientists determined the age of one clam to be 507 years old, making it the longest-lived non-colonial animal whose age has been accurately recorded.

335 years

The duration of the war between the Netherlands and the Isles of Scilly, from 1651 to 1986

This is the longest war ever, but no shots were fired and no lives were lost. The Netherlands allied itself with the English Parliamentarians against the last of the Royalists, who had been banished by Oliver

Giant tortoise

Ocean quahog

Cromwell to the Scilly Isles. The Netherlands declared war when the fighting had already ceased; the hostilities were forgotten and the war officially ended only in 1986.

The Isles of Scilly are located off the southwest tip of the English coast, near Cornwall. Of this group of 140 islands, only 5 are inhabited, with a total population of about 2,000. The islands are beautiful and are a favorite destination for nature walks and bird viewings.

351 years
The half-life of ^{249}Cf

Californium is an extremely rare chemical element, with atomic number 98. It does not exist in natural form, but can be created synthetically in a nuclear reactor. The element was discovered during an experiment at the University of Berkeley in California, hence its name. The isotope ^{249}Cf decays into ^{245}Cm, an isotope of the better known radioactive element curium. This happens through alpha-decay, a process where an alpha particle is emitted (alpha particles are identical to the nucleus of helium-4, with two protons and two neutrons). The process is as follows:

$$^{249}\text{Cf} \rightarrow {}^{245}\text{Cm} + {}^{4}\text{He}$$

Alpha radiation does not penetrate the skin deeply enough to cause any damage, as opposed to when it is inhaled or taken in via food.

432.2 years = 1.363×10^{10} seconds
The half-life of ^{241}Am

This chemical element was named after the continent of America, where it was discovered, analogous to the element europium. The atomic number of americium is 95 and it belongs to the actinides group, of which the better known elements plutonium and uranium are also part, with atomic numbers 94 and 92, respectively. All actinides are radioactive and therefore hazardous. Americium does not exist naturally on Earth, but can be produced in nuclear reactors by bombarding plutonium-239 with neutrons. After capturing two neutrons, ^{241}Pu is created, which disintegrates into ^{241}Am through beta decay. In other words, ^{241}Am has 95 protons and 146 neutrons in its nucleus. It has a reasonably long half-life of 432 years and is a known and hazardous waste product of nuclear reactors. It decays into neptunium.

^{241}Am is also able to cause a nuclear reaction and, in principle, can be used in nuclear weapons. However, the critical mass required for ^{241}Am to sustain such a reaction is about 60 kilograms, which is a lot more than is necessary for plutonium or uranium. This is why it would be much more difficult to fabricate a bomb using ^{241}Am. Nonetheless, the isotope is often applied in smoke detectors and radiography.

400 to 500 years
The duration of the 'short ice age'

Since the end of the last ice age, about 12,000 years ago, our climate has been relatively stable and constant. Through all sorts of methods, scientists attempt to measure the average temperature on Earth as a function of time as precisely as possible. There are known natural fluctuations, where the climate diverts from the average. ('Average' in this context usually refers to the average temperature measured in the period 1960–1990, just before global warming.) A known deviation from the mean is the so-called 'short ice age', which occurred mostly in the northern hemisphere, somewhere between 1400 and 1900.

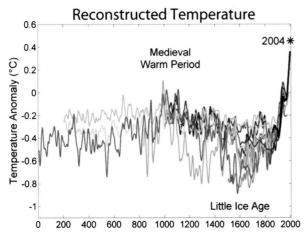

Diagram of temperature fluctuations of the past 2,000 years, until 2004. The different colors show the results of various scientific studies. There is a lot of disagreement regarding accuracy and interpretation of these results.

Measurements and research have indicated that the temperature of Western Europe averaged one to two degrees below the average value.

1,000 years, or thereabouts
Cycle of oceanic currents worldwide

Also relating to climate change, the circulation of water between the great oceans of the Earth, is known as the Global Conveyor Belt. Oceanic currents are caused by differences in temperature and concentrations of salt in the water, referred to as thermohaline circulation. 'Haline' is derived from the Greek word for salt, *halos*. The world map on the next page shows the direction of these worldwide currents. A part of this conveyor belt is the well-known Gulf Stream, a sea current of relatively warm water that flows from the Gulf of Mexico to Europe. The Gulf Stream keeps Europe warm, warmer than other areas on the same latitude, such as Canada, where winters are much colder.

The mechanism of the conveyor belt and the Gulf Stream is based partly on the concentration of salt. Saltwater has a higher density than freshwater; water with a lot of salt is thus heavier and drops to the bottom. Because of the warmer climate in the Gulf of Mexico, the water on the surface evaporates, increasing the salt concentration. This weighs it down, causing a convection current. A complicated system of aerial currents sends the water through the Atlantic Ocean in a north-easterly direction, until it meets Europe, where it transfers heat and cools down. The current then bends to the north where the surface water further decreases in temperature. This cooling surface water increases in weight, which in turn causes a large mass of cold water to drop to the bottom of the Arctic Ocean. Then, a deep current of cold water runs from north to south. Because colder or saltier water sinks locally, water is pushed to the surface elsewhere. This also happens in the Indian Ocean and the Pacific, as shown on the map. The water that surfaces is heated up slowly, crosses the ocean driven by the wind, evaporates, after which the salt density increases, and sinks to the bottom once again.

This creates a worldwide conveyor belt of currents: cold currents deep underwater and

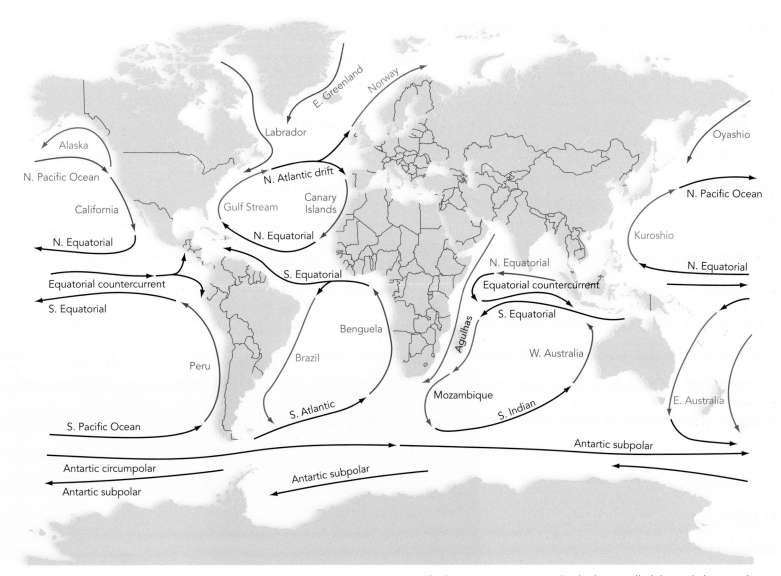

warmer currents on the surface. Scientists are busy studying these complicated phenomena to determine how changes in the cycle of the conveyor belt may influence climates. It is estimated that one full cycle takes about 1,000 years to complete.

The large water currents on Earth, the so-called thermohaline circulation. The position and direction of the currents depends on many variables, such as the direction of the wind, the Coriolis effect (the impact the spinning of the Earth has on currents), the concentration of salt, the temperature of the water, and the topography of the seabed.

10^{11} = 100 billion
10^{11} seconds = 3,171 years

Three thousand years brings us back to the ages before the beginning of the classical period of the great Greek and Roman civilizations, which began to flourish around 700 BC. Before then, Egypt and Mesopotamia reigned, but also the first Chinese dynasties and other ancient cultures existed. In about 3,300 BC, the first forms of script came into existence in Sumer, in what is now southeastern Iraq. Agriculture and irrigation already existed and the first cities arose. The Jewish people originated about 4,000 years ago. The oldest archeological finds from the city of David, the original Jerusalem, are dated to 3,700 BC.

2,000 light-years
The distance to a star where multiple planets have been discovered

The light we now see coming from the small star Kepler 11 was emitted about 2,000 years ago. The 'Kepler telescope' is actually a spacecraft (launched in March 2009) that is able to make extremely accurate observations of a group of about 100,000 stars. Its objective is to detect exoplanets, which do not belong to our Solar System. These can be spotted when they happen to pass in front of their parent star. Only after three occultations are we able to say that we have observed an exoplanet. We then also know its orbit. However, many planets do not pass directly in front of their parent star, which means we are unable to detect them, but

An impression of the solar system around Kepler 11, about 2,000 light-years away from our Solar System. The smallest of the six exoplanets shown is about 2.3 times heavier than the Earth.

because so many stars are being observed, we expect to find a plethora of exoplanets in the future.

The first exoplanets identified in this manner were located close to their parent star. As far as we know to date, starlet Kepler 11 has the most planets, with at least six. All are larger and heavier than the Earth, the smallest being 2.3 times as heavy. They must have a rocky composition, just like our planet. But because they are so close to their sun/parent star, it is too hot for any life as we know it to exist there. The search for exoplanets continues. In all likelihood more interesting planets will be found, perhaps closer than 2,000 light-years away and maybe even suitable for life.

4,600 years
The oldest pyramids

The first Egyptian pyramids emerged around 2,600 BC. The oldest pyramid — built from 2,631 to 2,612 BC is named after Djoser, a pharaoh from the third Egyptian dynasty who commissioned the build.

The large pyramid at Giza was completed around 2,560 BC and served as a tomb for pharaoh Khufu, also known as Cheops. Cheops was the second pharaoh of the fourth Egyptian dynasty. This pyramid is the only surviving Wonder of the World — the Classical era boasted seven wonders, although they are not all believed to have existed at the same time.

The Djoser pyramid took almost 20 years to build.

The Khufu pyramid (also known as the Pyramid of Cheops).

An ancient bristlecone pine. One of the oldest trees in the world, a bristlecone pine called Methuselah is more than 4,840 years old. The tree was named after the biblical figure Methuselah, who is believed to have lived for 969 years.

4,845 years
The age of the oldest living tree as of 2014

The age of a tree can be established by counting the rings of its trunk, though it is also possible to use carbon dating. At the end of this chapter we will explain how this works. Generally speaking, conifers get much older than deciduous trees. The oldest known conifer, baptized Prometheus, appeared to have been 4,844 years old. It lived in an American national park in Nevada, but was cut down for research in 1964 — they didn't know then that the tree was so old. The oldest living tree was then thought to be Methuselah, a bristlecone pine, that lives in the White Mountains of California and must now be 4,845 years old (as of 2014). However, a tree of the same species has been discovered that is now determined to be 5,064 years old and still alive. Its identity is kept secret. Some trees are claimed to be even older, 6,000 years or more, but because the earliest wood is now rotted away, exact ring counting or carbon dating is impossible, and the age can only be estimated.

5,000 Years
Dolmens in Northern Europe

A dolmen consists of a shelter built out of rocks, which covers a crypt or tomb to honor the dead. They belong to the oldest human inhabitants of northern Europe, people from the Neolithic Funnelbeaker culture of 3,400–2,900 BC.

Poulnabrone dolmen, County Clare, Ireland. Another example of a megalith from the New Stone Age is the well-known Stonehenge in England.

Funnel beakers were often found inside dolmens. In the Neolithic Age, also known as the New Stone Age, many objects and tools were made of stone. They were used for agriculture, which was just starting to develop. In this period, humans started to settle in one place, the wheel was invented for transportation purposes and the first scripts were created. After the Stone Age, the Bronze Age commenced (circa 3,000–800 BC), followed by the Iron Age (circa 1,000–0 BC).

5,730 years
The half-life of ^{14}C

Carbon — denoted by the letter C – forms the basis of organic chemistry. Life itself is based on chemical reactions between molecules in which the carbon atom plays a significant role. Some materials such as diamond and graphite consist almost exclusively of carbon. The element has atomic number 6, so ^{14}C contains six protons and eight neutrons. There are two stable isotopes, ^{12}C and ^{13}C, but ^{13}C hardly exists naturally on Earth. Unstable isotopes are, for example, ^{11}C with a half-life of 20.39 minutes, and the radioactive isotope ^{14}C, with a half-life of 5,730 years. ^{14}C is produced in the atmosphere, because cosmic rays can convert nitrogen into carbon. Half of the ^{14}C decays after 5,730 years into nitrogen by emitting a beta particle (an electron) and an anti-neutrino (see Chapter 4).

$$^{14}C \rightarrow {}^{14}N + e^- + \bar{v}_e$$

Carbon-14 is used for carbon dating, a technique to determine the age of certain organic materials. The method is based on the decay of ^{14}C; ^{14}C is built into living organisms, for example through photosynthesis in plants, which are then eaten by animals. But the manufacturing of this element ceases with the death of the organism, after which the built-in ^{14}C decays. By measuring the quantities of ^{14}C and ^{12}C after death, we can get an idea of when the organism died, and hence its age. An example of carbon dating is the determination of the age of the giant tortoise Adwaita (see Chapter 12). The ^{14}C-method is suitable for materials up to 45,000 years old. After many years, the concentration of ^{14}C has decreased to such a minimal amount that contamination with ^{14}C from later times would render any measurements useless.

Carbon dating measurements must be interpreted with care. What must be determined is: when did the organism stop inhaling carbon? We also have to establish the composition of the atmosphere at the time. A naive determination of the age of grass growing near a highway gives outcomes of tens of thousands of years. Of course that would be wrong; the grass has simply absorbed carbon that originated from oil obtained from fossils that were millions of years old.

Chapter 14

10^{12} = 1 trillion
10^{12} seconds = 31,710 years

Tera is the prefix that means 10^{12} or a thousand billion, or a trillion. So one terasecond is one trillion seconds. The word *tera* is derived from the Greek *teras*, which means monster. The timescale discussed in this chapter corresponds to the development of the first anatomically modern humans in Europe.

Homo sapiens sapiens came to Europe approximately 40,000 years ago. The Neanderthals already roamed this part of the world, but became extinct. It is believed that *Homo sapiens* migrated north from Africa; recent studies appear to indicate that certain *Homo sapiens* skulls are more than 195,000 years old.

32,760 years
The half-life of protactinium-231

At atomic number 91 in the periodical system, we find the chemical element protactinium, with its symbol Pa. There are 29 known isotopes, of which the most stable is ^{231}Pa, which halves in 32,760 years through the process of alpha radiation (helium nuclei) to ^{227}Ac or actinium. The element is named after this process: the Greek *protos* means 'first' in the sense of 'prior to'. Its isotope appears in the decay process of uranium-235, via thorium:

$$^{235}U \rightarrow {}^{231}Th + {}^{4}He$$
$$^{231}Th \rightarrow {}^{231}Pa + {}^{4}He$$

Protactinium is extremely poisonous and radioactive, and is very rare on Earth.

26,000 years
Precession movement of the Earth's axis

The Earth turns like a spinning top that does not quite stand up straight. Our planet is at an angle of 23.5 degrees to its orbital path. This angle is not quite constant; it varies a few degrees over time. As well as the spinning top motion that determines the length of our days, the Earth also makes a precession movement in the opposite direction to its spin around its axis (see drawing on page 62). In effect, the Earth's axis makes a cone-shaped movement.

The period of this precession is 26,000 years and it affects the beginning of spring and fall — the times when the Earth points neither towards nor away from the Sun (see Chapter 9). This means

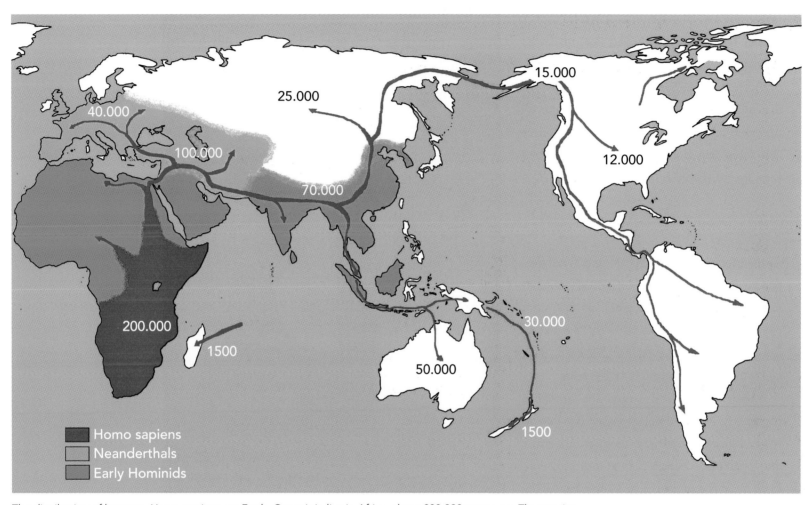

The distribution of humans, *Homo sapiens*, on Earth. Our origin lies in Africa, about 200,000 years ago. The species spread to Asia and Europe over time, where it was labeled *Homo sapiens sapiens*, a sub-species. Our ancestors often encountered another human species who lived there, the Neanderthals. In certain areas such as Australia, humans only set foot about 50,000 or 40,000 years ago, when Australia, New Guinea, Tasmania and a few other islands formed one continent. Around 30,000 years ago, during the Ice Age, the sea level was much lower, so that it was possible to walk from Australia to New Guinea. These areas became populated in this manner. Australia was first visited about 50,000 years ago by people from Asia, who almost certainly arrived over water. Not much is known about the relative sophistication of their ships, but this crossing is generally accepted as humankind's first great sea voyage. This period is often referred to as the "Great Leap Forward", because humans developed and spread rapidly during this time.

that spring and fall sometimes start a bit sooner or later. Combined with the fact that the Earth's orbit is not exactly circular, but rather elliptical, this means that sometimes summer lasts longer than winter. We are able to calculate that the astronomic summer on the northern hemisphere lasts seven days more than winter, but in 10,000 years this will be the other way around.

41,000 years
Vibration period of the Earth's axis

As mentioned above, the Earth's axis is not perpendicular to the Earth's orbit. At this moment in time, its angle is 23.44 degrees, but that is not constant. The angle varies between 22.1 and 24 degrees over a time period of 20,500 years. Back and forth, the total period is 41,000 years.

When the angle is larger, the northern hemisphere is tilted towards or away from the sun more, and consequently, the summers are warmer due to increased intensity of the solar radiation. In winter it is colder, because a larger part of the surface is turned away from the sun. When the angle of the tilt is smaller, the summers are cooler and the winters milder. It turns out that the two differences do not cancel out each other; the average temperature on Earth is lower at a smaller angle than with a larger inclination. Scientists believe, therefore, that longer periods of colder temperatures coincide with a smaller angular tilt of the Earth's axis and, in combination with other

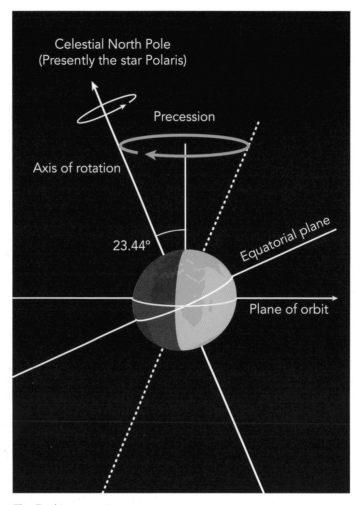

The Earth's precession movement

fluctuations of the Earth's orbital parameters, could be a possible explanation for the emergence of ice ages. At the moment the Earth's axis makes an angle of 23.44 degrees with its orbital plane and it is on its way to a smaller angle. It will reach the lowest angle in about 10,000 years from now, when we could expect another ice age.

2 × 10¹² seconds
70,000 years after the Big Bang

At this point in time, the universe has cooled down to 8,700 kelvins as a result of its expansion. At this temperature, particles move with an average energy of 0.75 electron volts (the energy unit of electron volt or eV is explained in Chapter 22) for each degree of freedom. Thus, free particles in a three-dimensional space, such as photons of heat radiation, have an average of 2.25 electron volts of kinetic energy. Due to the universe's expansion, photons continue to lose energy (this is known as 'redshift') and, because energy corresponds to mass, the strength of gravity caused by photons decreases. (Photons and rays of light do not have mass, but do have vibrational energy. According to Einstein's relativity theory, every source of energy, massless or not, creates a gravitational field.)

The moment has arrived when the force of gravity of matter is greater than the gravitational force of the existing radiation. This is a pivotal point in the history of the universe, because as gravitational forces are now mainly caused by matter, the situation becomes unstable: due to the so-called Jeans instability (see figure on the right), these particles are beginning to clot and form the first stars, solar systems and, much later on, planets. During this clotting process they follow the pattern of very small fluctuations that arose during the much earlier period of inflation of the universe. Indeed, the universe had been homogeneous during the time in between, but not *perfectly* homogeneous, so that the matter particles found seeds to begin their clotting process. The universe is about ten million light-years in size at this point and its density is about two million protons per cubic centimeter.

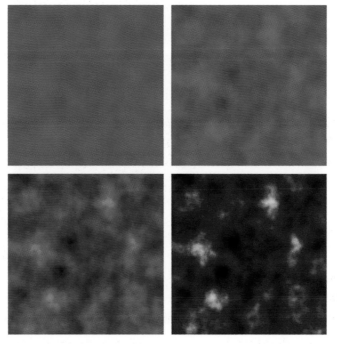

When gravity gets a grip on matter in the universe, it clusters. This occurs at every time interval, therefore small and large clustering occurs equally as fast. This is how galaxies, stars and planets come into existence.

Chapter 15

10^{13} = 10 trillion
10^{13} seconds = 317,098 years

In the preceding chapter, we looked at the migration of *Homo sapiens* to Europe and other parts of the world. The cradle of our species is found in Africa, where about 200,000 years ago it came into existence within the genus *homo*. This has been determined with the aid of fossil records and DNA research. *Homo sapiens* is the only remaining, living species of this genus. A timeline of the predecessors of *Homo sapiens* and other hominids, such as Neanderthals, will be shown in the next chapter. As its name suggests — *sapiens* in Latin means wise or rational — the *Homo sapiens* has a well-developed brain, which he uses to think and to communicate with fellow humans. This enables him to solve relatively easy problems. Because he stands up straight, he can use both hands to pick up objects and to construct tools.

Homo sapiens

245,500 years
The half-life of ^{234}U

Uranium has atomic number 92, which means the nucleus of isotope ^{234}U has 92 protons and 142 neutrons. It decays via alpha radiation into thorium. Everyone knows about uranium and its application in nuclear reactors and weapons, but it is isotope ^{235}U that is often used in this application as it is more common on Earth than the element ^{234}U. There is a large variety of uranium isotopes, and some of these — with an even longer half-life — we will meet in other chapters. The element is extremely poisonous and radioactive because of the large quantities of alpha radiation.

340,000 years
The half-life of ^{248}Cm

Curium is named after Marie and Pierre Curie, for their work on the phenomenon of radiation (see also Chapter 3). In small quantities — typically a microgram — curium can be reproduced by shooting alpha particles (helium nuclei) at plutonium. Curium disintegrates back into plutonium by the reverse process. ^{247}Cm is its most stable isotope, with a half-life of 15.6 million years, followed by ^{248}Cm with a half-life of 340,000 years. Its decay is expressed as follows:

$$^{248}\text{Cm} \rightarrow {}^{244}\text{Pu} + {}^{4}\text{He}$$

In total there are 19 known isotopes of curium and none of these is stable. Some have a half-life of three hours or even less.

Once per 300,000 years, on average
Reversion of the Earth's magnetic field

Our Earth has a magnetic field. Seafarers make use of it when referring to their compasses to determine a course. The needle of a compass points to the magnetic north pole, but this is not at the exact same location as the geographic north pole.

The magnetic axis of the Earth does not coincide with the rotational axis of the Earth, but is at an angle of 11.5 degrees to it. The magnetic axis is also not at a fixed position, but can actually turn around gradually, meaning the magnetic north pole becomes the magnetic south pole.

This can be measured in rocks with magnetic properties. They contain minerals that are oriented in the direction of the Earth's magnetic field, whatever that was at the moment they were formed. The reversal of the magnetic poles is probably caused by electric currents within the Earth's core. On average, during a time span of 150 million years, the Earth's magnetic field reverses every 300,000 years, but there are large fluctuations in the lengths of these periods. As such,

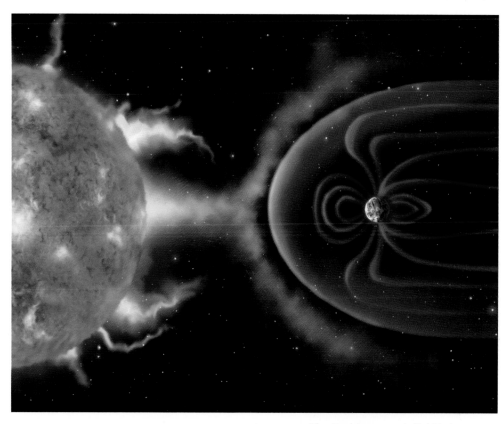

we cannot talk of a periodical signal. The current period, for example, is already twice as long as the average timeframe.

The strength of the magnetic field is expressed in the unit tesla, abbreviated to T. The Earth's magnetic field is not very strong and varies across its surface. Currently, a magnetic field of 3×10^{-5} T is measured at the equator. In comparison, the strength of the magnetic field near a bar magnet, such as you might find in a toy train, is 0.1 T. Near a mobile phone, the strength is 10^{-4} T or 0.0001 tesla.

The Earth's magnetic field is important because of its role in providing protection from solar winds — a stream of electrically-charged particles ejected from the upper atmosphere of the Sun that bombards our planet. This high-energy radiation is dangerous for life on Earth, but is mostly deflected by its magnetic field. Outside the relative safety of our atmosphere, space travelers must arm themselves against the solar wind. Measurements have indicated that the force of our magnetic field has decreased substantially over the last two centuries. That is a cause for concern, because a weak magnetic field offers less protection against the Sun's dangerous radiation.

1.2 × 10¹³ seconds
380,000 years after the Big Bang

The age of approximately 380,000 years marks another key turning point in the history of the universe. The temperature has fallen to 2,970 kelvins. At this temperature, electrons are being caught in atomic nuclei for the first time, so that the majority of atoms, mostly hydrogen and helium, are now electrically neutral. This process is called 'recombination' in physics. This leads to the important phenomenon that rays of light are now able to travel through the cosmos, as they are no longer blocked and scattered in the plasma of protons and electrons; thus, the universe is becoming transparent. The light rays that are emitted at this temperature are therefore still visible, even if they are — as a result of the expansion of the universe — cooled to a temperature of 2.725 degrees above absolute zero, more than a thousand times smaller (indeed, since then, the universe has expanded by more than a factor of 1,000 in each direction). This is the cosmic background radiation that is omnipresent. Recently, small fluctuations in radiation patterns have been detected, providing us with a wealth of information about the early stages of the universe, when it was 35 million light-years across and, with only 50,000 protons per cubic centimeter, already very diluted.

Cosmic background radiation, measured by the European Planck Satellite, a space telescope that maps cosmic background radiation very precisely. The map above shows us exactly what the universe looked like 380,000 years after the Big Bang, with lenses that see only microwaves. Take note of the small fluctuations in temperature: the small yellow and red areas are a little warmer than the green and blue ones. These variations probably emerged in a much earlier stage of the universe, right after the Big Bang. It is suspected that these fluctuations were actually much bigger, but they were ironed out by a period of rapid expansion, also called *inflation*. Cosmologists are busy developing a theory of inflation that explains these fluctuations (see Chapter 22). New observations shall provide us with fresh data about cosmic background radiation, and we hope to be able to understand more about this early stage of the universe in the next few years.

479,452 years

Traffic congestion per year in the United States

The Texas Transportation Institute (TTI), part of Texas A&M University, calculated how much time the average American spent in slow traffic in one year. In 2005, this number was 37.4 hours per driver. In 2007, this decreased to 36.1 hours, but that is still almost one working week per person per annum. Or worse, a week of holiday!

The population of the United States is about 300 million, but not all of these people drive, nor do they spend time in congested traffic every day. Research carried out by TTI shows that in 2007 a total delay of 4.2 billion hours had been caused by congestion. That is almost half a million years. We would not be surprised if these alarming numbers are similar in Europe.

Chapter 16

10^{14} = 100 trillion
10^{14} seconds = 3.17 million years

The first footprint of a human species, literally, is believed to date from 3.5 million years ago. As the result of continuous evolution, three new forms developed out of *Australopithecus afarensis* (a human species on two legs). Two died out quite quickly. The third kind, *Homo habilis* (the 'handy' man) must have lived around two million years ago and developed into *Homo erectus* (upright man). *Homo habilis* developed sharp stones into tools, for example as hammers or cutters, and launched us into the Stone Age. At present, we hardly use stone any more to manufacture tools, but turn to iron or other metals instead. The Stone Age lasted for about 2.5 million years and made way for the Bronze Age, followed by the Iron Age.

It is believed that in the period between 4 and 8 million years ago, several types of human species made an evolutionary split from apes. The ape most closely related to humans and still alive today is the chimpanzee, with DNA that is 98.5% identical to that of humans. One of these hominids was *Australopithecus*, which evolved into the genus *Homo* — the human. Within this genus, multiple species existed, the earliest being *Homo habilis*.

For a long time, it was believed that this species evolved into *Homo ergaster* (the 'workman') and eventually became *Homo sapiens*, the modern man (see figure on page 69). Another possibility is that the evolutionary paths of *Homo habilis* and *Homo*

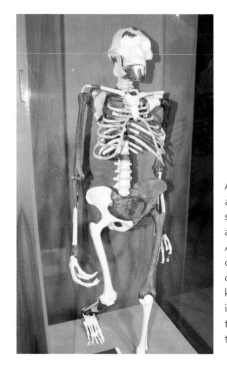

A reconstruction of *Australopithecus afarensis*, one of the first human species (*Australis* is Latin for 'Southern' and *pithekos* is Greek for 'monkey'). *Australopithecus* was about 120 centimeters tall, and he was able to walk on two legs. The oldest and most well-known fossil of *Australopithecus afarensis* is the female skeleton called Lucy, dated to 3.2 million years ago. It was found in the area of Afar in Ethiopia.

ergaster continued parallel to each other and the two species co-existed, and that each developed separately. Further research shall clarify the exact relationships between the various species of man.

2.6 million years
The duration of the Pleistocene period

The history of the Earth is divided into geological periods and epochs. We will discuss several aspects of these eras in more detail in the coming chapters. Here, we look at the Pleistocene, the epoch of 2.588 million years ago until the end of the Ice Age, about 11,700 years ago. It is the time of ice ages, or glacial periods, that were punctuated by more or less warm interglacial eras.

During the Pleistocene, dozens of ice ages appear to have taken place, during which the temperatures on Earth were considerably lower for thousands of years at a time. Glaciers in mountainous areas and the ice caps at the Earth's poles were much larger than they are now. Even parts of Western Europe, such as the Netherlands, were covered in snow or ice during some of these ice ages. During this period, ice age animals came into being, such as the mammoth, the woolly rhinoceros and the giant reindeer, which had antlers spanning a width of four meters. Because a large fraction of the existing water was fixed in the continental ice caps, the sea level was very low. In the Pleistocene, numerous land animals lived at the bottom of what are now shallow waters, such as the North Sea.

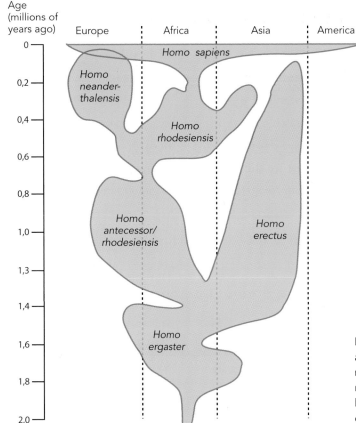

Simplified diagram of the evolution of the genus *Homo*, showing when and where the various species lived. *Homo habilis* is presumed to be the predecessor to *Homo ergaster*.

During the last Ice Age, animals such as the mammoth and the woolly rhinoceros roamed the landscape, including exposed tundra that now lies at the bottom of the North Sea.

The end of the last ice age, 11,700 years ago, brought about dramatic changes. The global climate heated up considerably within a few decades. The change in temperature caused a lot of the land ice to melt, creating the North Sea. Many animal species were not able to cope with the drastic change in temperature and died out. A new age dawned, the Holocene, an interglacial period that we still enjoy today. Hardy humans — evolved from *Homo habilis* to the modern man *Homo sapiens* — survived the whole Pleistocene, including its ice ages.

3 million years
The half-life of ^{154}Dy

Dysprosium has atomic number 66 and as such belongs to a group known as lanthanides, which are rare Earth metals. All lanthanides are silvery-white, relatively soft metals that burn easily, and also react strongly with water and hydrogen. Represented by the first element of the group, lanthanum, they also have other chemical characteristics in common.

The name dysprosium comes from the Greek *dysprositus*, which means 'difficult to isolate'. In fact, techniques were invented not long ago that do enable us to isolate the stable isotopes of the element from the mineral compounds in which it occurs. Dysprosium is used in lasers, and because of its magnetic qualities there are various other uses, such as in compact discs. It would have been possible to use it in hybrid electrical cars, if not for the inconvenient fact that there is too little of the mineral in the natural world. It has a high melting point, 1,680 kelvins, and can be used in nuclear installations that require materials that resist high temperatures. Of all unstable isotopes of dysprosium, ^{154}Dy has the longest half-life. It decays via alpha radiation into ^{150}Gd, an isotope of gadolinium.

3 million light-years
The distance to the Triangulum Galaxy

Our Sun is one of hundreds of billions of stars in the galaxy that we call the Milky Way. It is not known precisely how many stars make up our galaxy, but it is estimated to be between 200 and 400 billion. In addition, there are at least a hundred billion other galaxies of varying sizes in the Universe. The closest is Andromeda (M31), at a distance of about 2.5 million light-years. There are also smaller star groups around 180,000 light-years away — the Magellanic Clouds — but these are really satellite systems of our own Milky Way.

It might be that the Andromeda Galaxy contains more than a trillion stars. Also close by is M33, the Triangulum Galaxy, at 3 million light-years from Earth. This system is quite a bit smaller than the Milky Way and contains an estimated 30 to 40 billion stars. Through gravity, the Triangulum Galaxy is loosely connected to the much bigger

Andromeda Galaxy. M33 is located in the constellation of the Triangle and is sometimes visible on Earth with the naked eye. Together with the slightly brighter Andromeda Galaxy, these are the most distant objects that we can observe without telescopes. The Triangulum Galaxy has a spiral form, with a diameter of approximately 50,000 light-years. It belongs, together with the Milky Way, the Andromeda Galaxy and a few other smaller galaxies, to the so-called Local Group.

The Andromeda Galaxy, at a distance of 2.5 million light-years from Earth.

The Triangulum Galaxy, at 3 million light-years from Earth.

Chapter 17

$10^{15} = 1$ quadrillion
10^{15} seconds = 31.7 million years

In English, 10^{15} is a 'quadrillion'. In Dutch, this number is called 'biljard'. 10^{12} is called a 'billion' in most countries in continental Europe, while 10^9 is called a 'billion' in most English and Arabic-speaking countries. The variation in names for these large numbers is due to the fact that most continental European countries use what is called the 'long scale' naming system, and the UK and the US the so-called 'short scale' (to add to the confusion, the UK officially changed to the short scale in 1974, although even today many British people commonly favor the long scale).

Large numbers up to 10^9 have the same names in both systems, but beginning with 10^9, names for large numbers start to differ. In the short scale, each new term greater than a million is 1,000 times the previous terms. 'Billion', therefore, means a thousand millions (10^9). 'Trillion' means a thousand billions (10^{12}), a quadrillion is 10^{15}, and so on. In the long scale naming system, however, each new term greater than a million is 1,000,000 times the previous term: a billion means a million millions (10^{12}), a trillion a million billions (10^{18}), and so on. The short scale jumps with 3 powers of 10, while the long scale jumps with 6 powers of 10. We have summarized the long and short scales in a separate section after the introduction. Throughout this book we use the short scale, even though the original (written in Dutch) used the long scale.

Geological timescales

The continents' positions are not fixed. Over tens of millions of years they moved towards and away from each other. Around 65 million years ago Africa was separated from the continents of Europe and Asia, but over a period of 40 million years this changed. Africa bumped into Eurasia, creating the Alps. A connection was made between South and North America, and Australia broke away from Antarctica. India, which had split from Africa long before, continued to proceed with its journey over thousands of kilometers, to eventually join Asia. This caused the Himalayas to

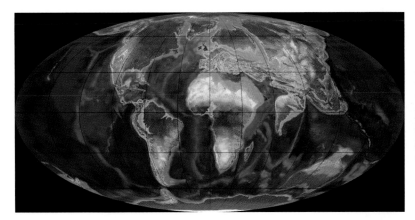

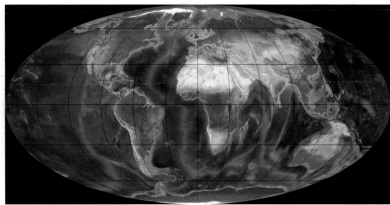

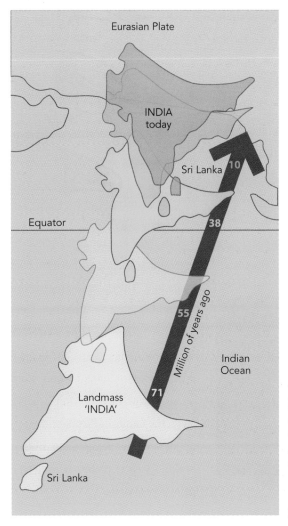

appear about 10 million years ago. This movement continues to this date; the Himalayas are rising by 5 millimeters each year.

The geological period during which these events happened was formerly called the Tertiary, the third long geological period in the history of the Earth, from about 65 million to 2.5 million years ago, when the Pleistocene period started. In modern paleontology, the Tertiary is divided into two sub-periods, the Paleogene (65 to 25 million years ago) and the Neogene (25 to 2.5 million years ago). The pictures above show the positions of the continents during these eras.

Just before the Tertiary, at the Cretaceous period (see next chapter), our atmosphere contained a lot more greenhouse gases and the temperature was much higher than it is today. But then the climate changed. Continental drift forced the oceans to expand and the water circulation changed. The warm sea currents along the equator were obstructed by the land bridge between South and North America, and by the attachment of Africa and Eurasia. This created

Top left: location of the continents in the Paleogene period. Top right: Position of the continents in the Neogene. During the Neogene, the Alps and the Himalayas appeared. Australia (from the Latin *australis*, which means 'South') moves towards the north. Research has indicated that Australia will bump into Asia in about 40 million years.

The separate landmass 'India' began its arduous journey across the Indian Ocean more than 70 million years ago, swallowing some islands on its way. By pushing against the Eurasian Plate, it created the Himalayas tens of millions of years ago.

73

colder currents flowing from the South Pole to the North Pole, cooling the Earth. Climates changed around the globe. It is also possible that the drastic climate change and diminished greenhouse gases at the start of the Paleogene were the result of a large meteor that hit the Mexican Yucatán peninsula at roughly the same time.

Life on Earth responded significantly to the rapid climate change. While dinosaurs ruled the world in the Cretaceous, they were succeeded by many new species of mammals in the Paleogene. Some of these animals are still around today, even if they look rather different after millions of years of evolution. Horses, for example, evolved from small, four-toed, grazing animals. To escape from predators, they developed stronger legs and learned to gallop. Over a period of 60 million years, they grew to about three times their original height, from half a meter to more than 1.5 meters, which is typical today.

Elephants evolved in a similar manner. Their original ancestors lived about 50 million years ago, when they were part of the family known as ungulates.

Moeritherium, an ancestor of the elephant.

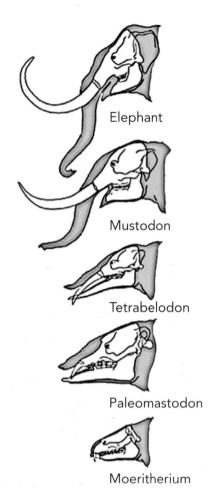

Elephant

Mustodon

Tetrabelodon

Paleomastodon

Moeritherium

(ca. 5 million years ago – today)

(ca. 5 million years ago)

(ca. 20 million years ago)

(ca. 30 million years ago)

(ca. 37 million years ago)

Fossils show that the first animals to boast a trunk-like appendage, the Moeritherium, lived about 37 million years ago. They already had a small trunk — more like a long nose — and they were believed to have lived partly in or near the water, like an amphibian. In fact, their name means 'the beast from Lake Moeris', which was an ancient water source that existed in what is now Egypt. After a series of evolutionary steps, the family of Elephantidae arose from this genus about five million years ago. Both extinct mammoths and modern elephants belong to this genus (see the drawings above).

The evolution of the elephant, from a large snouted swamp dweller of 40 million years ago, to one of the first real elephant-like animals, complete with a trunk, almost five million years ago. Note the proportions of the skull compared to the soft nose parts.

1.1×10^{15} seconds = 34.7 million years
The half-life of ^{92}Nb

Niobium is a chemical element with atomic number 41. It does not exist in a pure form in nature, but in compounds in several minerals. Being a transition metal, niobium is used to enhance the electric conductivity of materials. The element itself in its pure form (elements can be purified in laboratories) is superconducting at 9.3 kelvins, the highest temperature of all superconducting elements.

There are about 25 known isotopes of niobium, and only one of them, ^{93}Nb, is stable. With one neutron less — 51 neutrons — we arrive at ^{92}Nb, with a half-life of 34.7 million years. Via beta decay this isotope decays into molybdenum.

60 million light-years
Roughly the distance to the Virgo cluster

Stars belong to galaxies, such as our Milky Way or the Andromeda Galaxy. In our universe, most galaxies are part of a group or a cluster. For example, the Milky Way, the Andromeda Galaxy and the Triangulum Galaxy, (see the previous chapter) together with another 30 or so smaller galaxies, belong to the so-called Local Group. The diameter of the Local Group is about 10 million light-years.

In the landscape of clusters, Virgo is our largest neighbor — at a distance of roughly 60 million light years from Earth. It is positioned in the constellation Virgo, hence the name, and contains almost 2,000 galaxies. As such, the cluster is much larger and heavier than the Local Group. The Virgo cluster's mass is about 10^{15} (or one quadrillion) solar masses. At this moment, the Local Group is moving away from the Virgo cluster at 1,000 kilometers a second. However, because of the immense gravitational force of the Virgo cluster, the Local Group's speed will decrease, and will eventually be absorbed by Virgo.

The 'center' of the Virgo cluster. The black dots, in the foreground of this picture, are stars in our own Milky Way, which have been redacted to illuminate the extra-galactic nebula.

Chapter 18

10^{16} = 10 quadrillion
10^{16} seconds = 317 million years

Our Earth has changed significantly over the last 10 quadrillion seconds. More than 200 million years ago, all continents were connected, forming a giant continent, Pangaea — after the Greek 'all' (*pan*) and 'earth' (*gaia*). Large parts of the supercontinent were desert-like and unsuitable for any life. As a result of plate tectonics, Pangaea broke up into smaller pieces and became the continents we know today. As more land came into contact with water, the biodiversity of the Earth blossomed.

180 million years

Era of the dinosaurs

For about 180 million years, between 245 and 65 million years ago, dinosaurs roamed the Earth. It was the age of reptiles, including dinosaurs, but also the period during which birds first took flight, being direct descendants of dinosaurs.

The first birds, such as *Archaeopteryx*, still had teeth and a long tail and greatly resembled the feathered bipedal dinosaurs known as theropods. A lot of discussion remains today as to the origin of flight: flying could have originated from the ground,

The *Brachiosaurus*. This was one of the largest and heaviest dinosaurs that ever lived. It could be up to 23 meters long and 12 meters high, weighing 50 tons.

by running increasingly fast with longer steps that turned into jumps; or from above ground, by 'gliding' out of treetops. Perhaps it was a combination of the two.

Dinosaurs became extinct at the beginning of the Paleogene (see previous chapter), probably because they were unable to adapt quickly enough to a climate that changed rapidly as a result of a large meteor hit in Yucatán.

220 million years
Orbital period of the Sun, together with all its planets, around the center of the Milky Way

Our Sun is only one of many stars in our galaxy, the Milky Way. With the naked eye we are able to observe thousands of stars, but research has shown that the Milky Way actually contains far more, between 200 and 400 billion. In the 18th century, William Herschel discovered that the Sun is part of a 'disc' of stars. Many galaxies in the universe have a spiral profile, with several winding arms encircling the center. The Milky Way has a diameter of about 100,000 light-years, and a thickness of about 6,000 light-years. The Dutchman Jacobus Cornelius Kapteyn (1851–1922) was the first astronomer to show that the Milky Way rotates around a central point. Both Herschel as well as Kapteyn placed the Sun at the center of the Milky Way. However, we now know that the Sun is about 32,000 light-years away from the center of our galaxy.

There are strong indications that the center of our galaxy contains a black hole, with a mass of two to three million times that of the Sun. In some other galaxies it is believed that much larger black holes exist, which could be billions of times as heavy as our Sun.

The Milky Way does not rotate as a rigid object; the rotational period at the center is much shorter than that of objects on the periphery. Calculations based on the motion of stars and the distribution

Reconstruction of the Milky Way, seen from above. In fact, we are right in the middle of one of its concentric rings, meaning we can see billions of stars of our galaxy as a strip on the horizon.

77

of their masses show that the rotational period of the Milky Way, where our Sun is located, is about 200 million years. The age of the Sun is about five billion years (see Chapter 19) and calculations show that it is about halfway through its life cycle. In its entire lifetime, the Sun will orbit 50 times around the center of the Milky Way.

**400 million years after the Big Bang
Formation of the first stars in the universe**

The 'dark age' of our universe lasted for 400 million years after its creation. Little about this period can be observed, but we know that the first stars come into being, so-called population III stars. Many of them were very heavy and exploded fast, as supernovae. Nuclear reactions in the interiors of stars created the first heavy elements, and it is due to these first violent explosions that heavy elements began to appear in gas and dust clouds. These clouds were hot, so that partial reionization of neutral hydrogen occurred.

The universe was already a billion light-years in size and its average density was three protons per cubic centimeter.

**703.8 million years
The half-life of ^{235}U**

Everyone has heard of uranium in the context of nuclear reactors or weapons, such as certain atomic bombs. The element is named after the planet Uranus, which had been discovered only a few years earlier, in 1781. The two most important isotopes are ^{235}U and ^{238}U, but ^{235}U hardly exists in nature. About 99.28% of natural uranium is ^{238}U, 0.7% is ^{235}U and an even smaller fraction is ^{234}U. Natural uranium must be enriched in laboratories to contain at least 3% ^{235}U for use in nuclear reactors, and at least 90% for nuclear weapons. Both isotopes have enormously long half-lives, 703.8 million years and 4.468 billion years (see also the next chapter). They are therefore used to determine the age of the Earth, through radio dating (a technique similar to carbon dating). Uranium-235 decays via alpha emission into thorium:

$$^{235}U \rightarrow {}^{231}Th + {}^{4}He$$

Both nuclear reactors as well as nuclear weapons are based on the principle of nuclear fission. This process splits the atomic nucleus into two by bombarding it with neutrons. The nuclei of only a handful of known atoms — ^{235}U was the first one — can be split. In the case of ^{235}U, by shooting at it with a neutron, the nucleus is split into two chunks, plus two or three neutrons and a large quantity of energy. Fission products vary strongly, but the most common ones are barium and krypton, as shown in the drawing above. The neutrons that are released as a result of this process may cause fission in other ^{235}U atoms, creating a chain reaction. In this manner, enormous quantities of energy can be released.

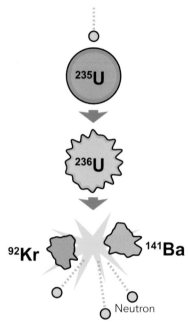

Fission in uranium-235.

Chapter 19

10^{17} = 100 quadrillion
10^{17} seconds = 3.17 billion years

With a panorama over billions of years, we leave behind geological periods and the evolution of life, and witness the maturing of the Earth as a planet. It took our world a few billion years to become what it is today. At 3.17 billion years in the past, we find ourselves in the midst of astronomic and cosmological developments, such as the emergence of solar systems, galaxies and stars, and, somewhat longer ago, the origin of the universe itself. That said, some events on an atomic scale occur in this timeframe as well, which we will discuss in more detail below.

3 to 3.5 billion years ago
The origin of life on Earth

Exactly how and when the first living organisms appeared on planet Earth is not known. They left no fossilized traces, and furthermore, most rocks on Earth are not old enough to show any recognizable evidence. When the Earth was formed, it had an atmosphere that would have been very hostile to modern life forms. It contained methane, carbon dioxide and other toxic gases, but no free oxygen. This 'Hadean Earth' was constantly bombarded by large meteors. Intense cosmic radiation and lightning may have created organic but lifeless material called the 'primordial soup'. The Moon, much closer to the Earth than it is today, must have generated intense

Representation of the creation of the solar system.

tidal currents in the primitive oceans. There are many theories as to how these conditions, in some ponds or deep under water, may have allowed the first self-replicating molecules to form. They were the forebears of life. The laws of evolution dictated their subsequent development. Around 2.5 billion years ago, sizable amounts of free oxygen appeared in the atmosphere, due to photosynthesis. It was during the 'Cambrian explosion', some 600 million years ago, that life began to proliferate and oxygen levels came close to today's values.

4.54 billion years
The age of the Earth

The age of our planet, about 4.54 billion years, has been determined with a margin of error of a mere 1%. Scientists have deduced this through the dating of the oldest known minerals and rocks. And from dating rocks from the Moon, it has been determined that it is about the same age. In fact, the Sun, Earth and all other planets in our solar system were created at the same time, from a whirling cloud of cosmic gas and dust. This cosmic nebula consisted mainly of light elements such as hydrogen. Heavier elements were carried towards us by exploding stars in the neighborhood, so-called supernovae. These would later prove essential for the formation of planets, and most certainly for the creation of life on one of them. The influence of gravity increased the density at the center of this maelstrom. This is how the Sun was created, at the center of our solar system, about 4.6 billion years ago. The remaining particles clustered

together to form chunks, which continued to develop into planets and their moons.

Our young planet Earth was very hot in the beginning, and it took about 100 million years for it to solidify. Numerous volcanic eruptions caused gases to escape and eventually form the Earth's atmosphere. Originally there was no water on the planet, just vapor, dissolved in molten rock under the Earth's crust. Through the cooling of the Earth and the volcanic eruptions, these vapors escaped and condensed, creating oceans. It is assumed that additional water was provided by ice comets, increasing the oceans' volumes. The exact explanation of the origin of water on our planet — and why there is no or little water on other planets — is much discussed and researched.

4.468 billion years
The half-life of ^{238}U

In the previous chapter, we covered uranium isotope ^{235}U, with its fissionable atomic nucleus. Although ^{238}U is not fissionable, bombarding the isotope with neutrons creates plutonium-239 via two processes of beta decay. In contrast, the nucleus of ^{239}Pu is fissionable and this isotope is therefore used for nuclear energy and, predominantly, nuclear weapons. Via alpha decay, ^{238}U becomes thorium-234.

14.05 billion years
The half-life of ^{232}Tu

Thorium-232 has 90 protons and 142 neutrons. It is the most stable and common isotope of this

A collision of two galaxies. Tidal forces bring about the fusion of two spiral systems to form a single elliptical galaxy.

very heavy element and could be the energy source of the future. It is technically possible to harvest nuclear energy from ^{232}Tu by shooting protons at it from a particle accelerator. As opposed to uranium, relatively few hazardous radioactive waste products are created through this process. The world stock of thorium is very large — it would not be depleted for thousands of years. A large proportion of the heat production at the center of the Earth is attributed to thorium and uranium decay. Thorium-232 eventually decays by emitting an alpha particle (helium nucleus), after which a series of radioactive isotopes decay rapidly, ending with the stable lead isotope, ^{208}Pb.

Five billion years from now
Collision between the Milky Way and the Andromeda Galaxy

The Andromeda galaxy is about 2.5 million light-years from the Milky Way (or 2.4×10^{19} kilometers). Of all the larger galaxies, it is our closest neighbor (see Chapter 16). Galaxies are not stagnant, but move around in relation to each other. The Andromeda galaxy is rushing towards us at a speed of about 100 kilometers per second (360,000 kilometers per hour), meaning it travels roughly three billion kilometers per year in our direction. This might appear strange, considering the fact that in a continuous expansion of the universe, all galaxies are moving away from each other. However, the Milky Way and the Andromeda Galaxy are an exception, because they are under each other's gravitational influence. Initially the two were moving away from each other, but because the attraction was so strong, they are now closing in on each other. Whether they will really collide is not certain, as the speed of Andromeda's lateral movement is not precisely known.

10 billion years
The lifespan of a star such as the Sun

Stars are burning bombs. They radiate energy and light through nuclear reactions occurring deep in their interior. This radiation from our star, our Sun, provides us with light and warmth on Earth.

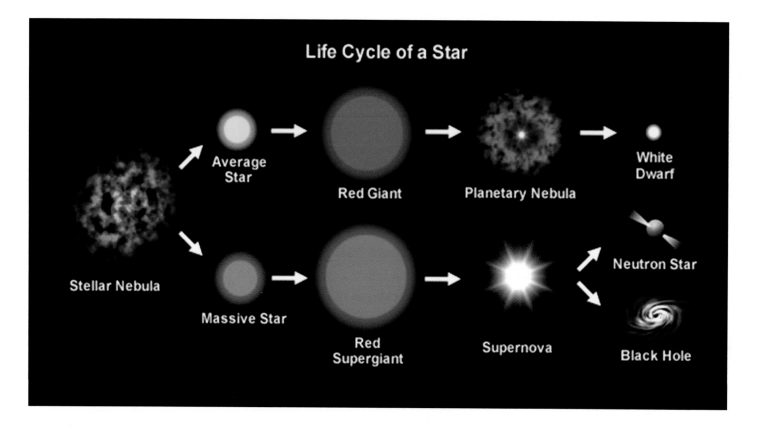

The outward pressure created by the star's high temperature is a sufficient counterforce to its own gravitational pull, ensuring it remains in balance. Towards the end of its lifetime, the fuel in the center is depleted, so that no further nuclear fusion reactions occur. The star still continues to radiate energy and will destabilize, creating an imbalance that causes it to implode under its own gravitational force. During this process the pressure and temperature rise again, so that nuclear fusion will now also occur in the outer layer of the star. This releases such a high amount of energy that the star will swell and become a so-called 'red giant'. When this happens to our Sun, its surface will come so frighteningly close to the Earth, that it will probably mean the end of our planet.

If the original star is much heavier than the Sun, it can explode — a phenomenon we call a supernova. A lighter star, such as the Sun, cools off after the red giant phase, and the inner layers contract, causing new types of nuclear fusion. Its final destiny is that of a white or brown dwarf, which consists mainly of carbon and oxygen. For heavier stars, of around

1.4 solar masses or more, the last phase is that of a neutron star or a black hole. Thus, the lifespan of a star depends partly on its mass; the larger the mass, the larger the inward pressure and higher the temperature, which accelerates the nuclear fusion. A star such as our Sun will live for about 10 billion years. That means it will extinguish in five billion years from now. A star of 10 solar masses will live for a much shorter time, about 30 million years. Smaller and lighter stars, of about 0.1 solar mass, have a very long life — in the order of 10^{12} years (1,000 billion years).

13.7 billion years after the Big Bang: today

The universe is now 13.8 billion years old, or 4.354×10^{17} seconds. In most regions of the universe temperatures have decreased to 2.725 kelvins above absolute zero. The most distant galaxies that we are still able to observe emitted their light when the universe was much younger, about 13 billion years ago. At that time, the galaxies were much closer together, but because of the rapid expansion of the universe, this distance has presently increased to almost 50 billion light-years. The average density today is 20 protons per cubic meter.

13.8 billion years ago
The Big Bang

We cannot go back any further in time than 13.8 billion years, or 4.354×10^{17} seconds; we have arrived at the moment of the Big Bang, the beginning of our universe. But what happened before the Big Bang? What did the universe look like then? These

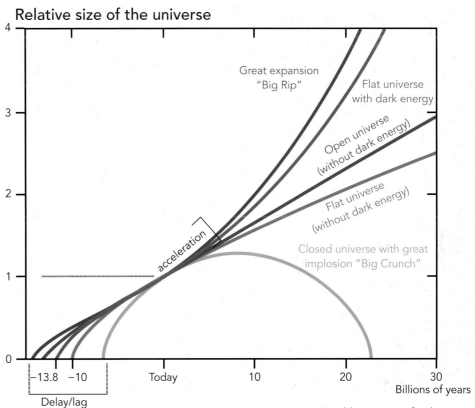

Relative size of the universe

Possible scenarios for the development of the universe. A flat universe with dark energy (the red curve) is considered by experts to be the most likely one. The expansion rate today is accurately known, which is why the curves all have the same slope. Thus it is found that the universe is 13.8 billion years old.

fundamental questions continue to fascinate scientists. Presumably, the notions of space and time cease to exist before the Big Bang, and the question about what happened beforehand makes as much sense as the question of what is north of the north pole. Our current understanding — or rather belief — is that the notion of time itself was created during the Big Bang, though various different scenarios have been proposed by researchers.

A Planck unit of time (or 10^{-44} seconds) after the Big Bang, the universe was about 10^{-33} centimeters in size (the Planck unit of length), much smaller even than an atomic nucleus (the diameter of an atomic nucleus such as hydrogen is about 10^{-13} centimeters. That is also about the size of a proton). Einstein's theory about gravity — general relativity theory – is not applicable to the Planck unit of time, because the theory must be combined with the principles of quantum mechanics of atoms and other particles in a so-called quantum gravity theory, but it has not been possible to do so yet. The search for such a theory continues at present, and is one of the most compelling unresolved issues in theoretical physics today, even though ideas do exist: a commonly studied model is the so-called string theory, see also Chapter 22.

Quantum gravity is to provide us with answers about the quantum universe, as it looked 10^{-44} seconds after the Big Bang. We will discuss this in more detail in Chapter 22. First, we discuss timescales that are much longer than the age of the universe; these mostly relate to our long and distant future.

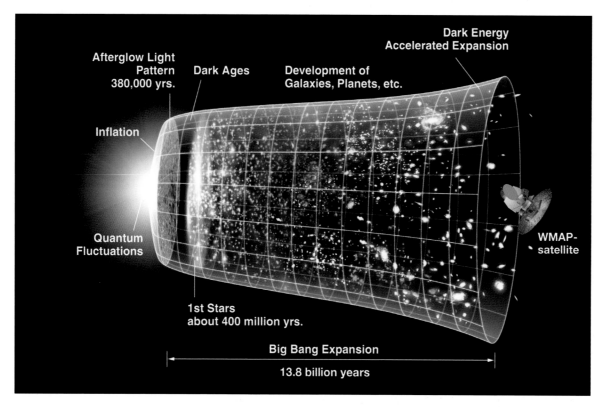

Simplified impression of the Big Bang and the expanding universe.

Chapter 20

The Large Timescales
10^{18} = 1 quintillion
10^{18} seconds = 31.7 billion years

10^{18} is also referred to as a quintillion, or one million to the third power. One quintillion seconds is also 3.17×10^{10} years, or 31.7 gigayears. This is a time span that is very difficult to imagine, but still, interesting things happen during these large timescales.

Perhaps you thought that the age of the universe would be the longest possible timescale? But then you would be limiting yourself to the duration of the past. There are various events and processes that take much longer than the present age of the universe, longer even than 13.8 billion years. Obviously these phenomena have not yet occurred, but nonetheless scientists are able to calculate the timescales, or at least make educated guesses about the order of magnitudes in time.

In this chapter we discuss a few examples, varying from certain radioactive isotopes with extremely long half-lives, to cosmological events that will happen at some point in the far future of the universe. Eventually we find ourselves in the 'dark eternities', which last quadrillions of times the age of the universe. What will the universe look like then?

We have included an overview of nomenclature concerning these enormously large numbers, see page xviii.

33.8 billion years
The half-life of the isotope ^{176}Lu

At the beginning of the 20th century, the French chemist Georges Urbain discovered the element with atomic number 71, which would later be given the name lutetium — *Lutetia* is the Latin name for Paris. The element exists only in very small quantities in the Earth's crust. Because isolating lutetium is costly, there are only a handful of applications in industry; for example, it is used as a catalyst in oil refineries.

^{175}Lu is the only stable one, but 30 or so unstable isotopes are known, with half-lives varying from 3.31 years (^{174}Lu) to less than a second. The major exception is isotope ^{176}Lu, with a half-life of

33.8 billion years. It is therefore the only unstable lutetium isotope that still exists in its natural form on Earth today. Because of its long half-life, ^{176}Lu is also used to determine the age of meteorites through radiometric dating. Most meteorites are about 4.5 billion years old, the same age as our solar system. Studying meteorites is important for gathering more insight into the origins of our solar system. We can learn more about the structure and properties of the existing matter in the early solar system from the composition of meteorites; the decay of relatively stable elements within meteorites can tell us their age.

30 billion years

The minimum lifespan of our universe

What does the future hold for our universe? After the Big Bang, it started to expand and our astronomic observations show that this process continues today. In fact, the expansion is accelerating. Will this continue forever or will the universe be brought to a halt at some point? If it stops expanding, the universe will contract as a result of gravitational forces — all galaxies will move towards each other and compress: the universe implodes.

This scenario is also referred to as the 'Big Crunch'. The universe will cease to exist, but it is also possible that a new universe will be created with another Big Bang. In this scenario, the universe could be cyclic, and it might even be that our Big Bang, 13.8 billion years ago, was not the first!

A closely related but separate question is whether the universe is spatially finite or bounded. Space in our universe may be curved, which is partly responsible for its evolution, together with existing matter and dark energy. If this curvature is negative, then the universe is spatially infinite, and we refer to this as an 'open universe'. If the curvature is positive, then the universe resembles the surface of a globe and would be referred to as 'closed'. In between these two scenarios is the possibility that the universe is flat and the curvature is zero; in this case the universe may be either infinite, or rolled up like a newspaper and thus finite.

To date, observations suggest that while locally there is some spatial curvature, it averages out to zero globally, just like a piece of paper that has bumps but is, on the whole, flat.

Dark energy, which is the cause of the universe accelerating, might prove to be mercurial. Most scientists believe therefore that the universe will continue to expand *ad infinitum*. Uncertainties and margins of error exist within the astronomic observations, but even allowing for the maximum discrepancy, we can deduce that the minimum lifespan of the universe is approximately 30 billion years. The maximum lifespan, on the other hand, may well be infinite.

The future of our universe. An open universe that expands forever, as in the picture, or a closed universe that will end in a Big Crunch?

The Large Timescales
10^{21} = 1 sextillion
10^{21} seconds = 3.17×10^{13} years

We are jumping ahead with a factor 1,000 — 3 powers of 10 — and arrive at one sextillion seconds, or 31.7 trillion years. It is the lifespan of the longest living stars.

10^{13} years

The maximum lifespan of a red dwarf

In Chapter 19, we discussed the lifespan of the stars and saw that our Sun can expect to reach the grand old age of about 10 billion years. Heavier stars live shorter lives because their nuclear fusion process is faster and their fuel depletes quickly. Light stars live longest, as they are very economical with their energy. Their temperature is much lower than that of the Sun and so they radiate weakly in reddish light. This is why they are called red dwarfs. Their mass is at most 40% of the solar mass, but often they are much lighter, with a mass of at least 7.5% of the solar mass. Because they radiate so weakly, they can carry on for a very long time. The smallest red dwarfs burn all their hydrogen and helium in a period of 10^{13} years. Of course, this point in time lies far off in a future we know little about — maybe the universe will 'crunch' before the first small red dwarfs peter out. It is likely that they end up as cooled down brown or black dwarfs, which no longer radiate. About 60% of all stars in the Milky Way consist of red dwarfs. A well-known red dwarf is Proxima Centauri — the star closest to the Sun — which forms a triumvirate along with the much brighter double star Alpha Centauri and Beta Centauri (see also Chapter 10). Proxima Centauri is not visible to the naked eye.

5.6×10^{13} years
The half-life of ^{184}Os

Osmium has atomic number 76, which means that ^{184}Os has 108 neutrons. It is a bluish-grey transition metal that is used in many alloys, such as the tip of a ballpoint pen, the needle of an old-fashioned record player, and some electrical connections. What is remarkable about osmium, is its extreme density: one cubic meter weighs 22,610 kilograms. After iridium (with 22,650 kg/m^3), osmium has the highest density of all chemical elements. The half-life of ^{184}Os is 5.6×10^{13} years, as it disintegrates via alpha decay into the better-known element tungsten, represented as ^{180}W.

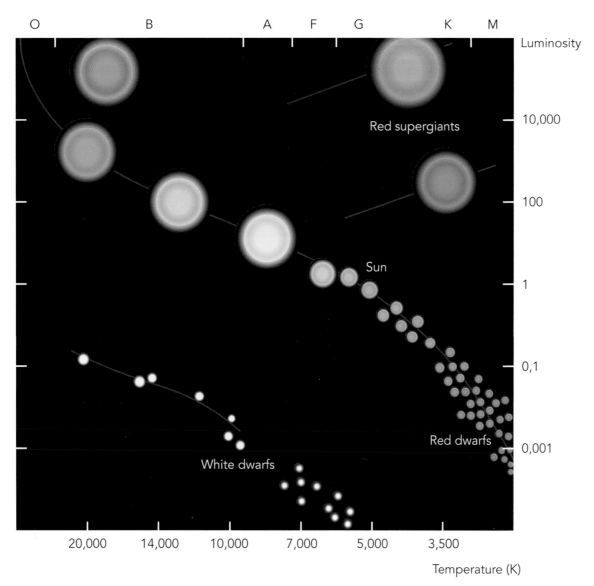

O   B   A   F   G   K   M   Luminosity

10,000

100

Red supergiants

Sun

1

0,1

Red dwarfs

0,001

White dwarfs

20,000   14,000   10,000   7,000   5,000   3,500

Temperature (K)

The Hertzsprung–Russel diagram is a graph that shows the absolute brightness of stars charted against the spectral type, which is really an indication of its temperature. During its lifetime, a star moves through this trajectory, though the exact path depends on the star's mass. Stars that are found at roughly the same position as others have a similar mass as well as age. Their remaining lifespan can also be estimated with some precision. First, all hydrogen is converted into helium; then the interior region of the star contracts, density and temperature there will increase, so that the helium is subsequently converted into heavier elements.

The Sun is found in the middle of the main sequence.

At the bottom right-hand corner, we find red dwarfs, such as Proxima Centauri, with extremely long lifespans of up to 10^{13} years.

It is amazing that we are able to determine such extremely long half-lives with any precision, considering these timescales are much longer than the age of the universe itself. Do not forget though, that we will not have to wait 56 trillion years before we can determine whether half of the unstable particles have indeed decayed. As we explained in the introduction, the decay curve is exponential. If we know a small part of the curve, we are able to predict its remainder. In other words, it is sufficient to start with a large number of particles and wait until a few have decayed. In the case of ^{184}Os, for instance, we can calculate that if we start with 10^{15} particles, 70 will have decayed after 5.6 years. It is not that difficult to gather this number of particles to monitor; in comparison, the number of hydrogen atoms in one gram equals the number of Avogadro, or 6×10^{23}. The weight of 10^{15} osmium atoms is only 0.3 micrograms. We are able to measure osmium's decay with sensitive detection devices.

10^{23} seconds (3×10^{15} years)

Our planetary system will cease to exist

How long will planets continue to orbit the Sun? Earth will, in all likelihood, be absorbed by the Sun long before we reach this very late era, if our star grows into a red giant as expected (see Chapter 19). Planets further away from the Sun, such as Jupiter, Saturn and others, may still exist. On very long timescales, planetary orbits are not stable and, moreover, the orbits of stars in our galaxy are irregular; it is also likely that every now and then a star will pass close to us, interfering drastically with our orbits. While these events are likelihoods and not certainties, which make them difficult to forecast, rough predictions and estimations can be made. Planets shall either get detached from the Sun or shall move into closer orbits, eventually being absorbed into the Sun after living for billions of years.

Osmium crystals

The Large Timescales
10^{28} seconds = 3.17×10^{20} years

As you may remember, stable isotopes live forever. But how far can we go with the half-lives of isotopes? Which are the longest living, unstable isotopes and how do they decay? This question brings us to timeframes of 10^{20}–10^{24} years. During this period, a new phenomenon arises, that of double beta decay. Some aspects of double beta decay are not yet understood, but perhaps new and interesting discoveries will change our understanding of the Standard Model of elementary particles. These might then also modify our present knowledge of double beta decay.

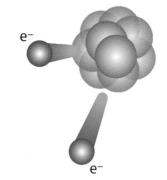

Double beta decay (left), and on the right the more unusual neutrinoless double beta decay.

1.08×10^{20} years
The half-life of ^{82}Se

Selenium has the atomic number 34. Its isotope ^{82}Se has 34 protons and 48 neutrons in its atomic nucleus. There are five stable isotopes, with 40, 42, 43, 44 and 46 neutrons, and more than 20 unstable isotopes, of which ^{82}Se has a half-life of 10^{20} years. Its disintegration is different from the common beta and alpha decaying processes, which we have discussed previously. Selenium-82 decays into krypton-82 via so-called double beta decay:

$$^{82}\text{Se} \rightarrow {}^{82}\text{Kr} + 2e^- + 2\overline{\nu}_e$$

Krypton has atomic number 36, so in this decay process two neutrons are converted into two protons. The decay is analogue to (single) beta decay, but the process converts the neutrons in a single process, whereby two electrons and two anti-neutrinos are emitted. If a single beta decay process could take place, ^{82}Se would then decay into ^{82}Br, but that is impossible, as the difference in mass between ^{82}Se and ^{82}Br is smaller than the electron's mass, meaning there is not enough energy to emit the electron.

Double beta decay has been observed in 10 elements, all with similar half-lives. This process is shown in the diagram above.

Double beta decay as pictured above (left) is rare, but completely natural and fits within our current understanding of elementary particle physics. Much more interesting is the process pictured on the right. Here, anti-neutrinos do not

90

escape from the atomic nucleus, but bump into and annihilate each other. When this happens, we refer to it as neutrinoless double beta decay. This process has never actually been observed, but is theoretically possible if a neutrino is its own anti-particle.

Fermions that are their own anti-particles are called Majorana fermions. No known elementary fermionic particle is a Majorana fermion, but it may be that neutrinos are an exception. This means that any observation of neutrinoless double beta decay would be fundamentally important for the formulation of the Standard Model of elementary particles.

Moreover, we have learnt recently that neutrinos have a very small mass, even if we do not yet know exactly how big it is. From determining the characteristic half-life of neutrinoless double beta decay — if it exists — we would be able to determine neutrinos' mass. Because neutrinos exist everywhere in the universe, they make an important contribution to the total matter-energy density of the universe, which in turn is a determining factor for the expansion or contraction of the universe.

3×10^{30} seconds = 10^{23} years
Instability of the earth's orbit because of gravitational radiation

By this point in time, the Earth will have long ceased to exist. The other planets will also have moved from their erstwhile orbits as a result of all kinds of chaotic disturbances. But even without turmoil on this scale — if for example only one lonely planet continues to orbit a star far, far away from any other stars — even this would come to an end at some point. A planet orbiting a (long since extinguished) star loses kinetic energy because of the inevitable gravitational waves it emits. We can calculate how long it will take for a planet — such as Earth, which orbits the Sun — to lose so much energy that it plummets into its star: this would take approximately 10^{23} years. The end would come quickly, because the intensity of those gravitational waves will increase with its increasing velocity.

And finally, we describe the isotope with the longest known half-life (also through double beta decay) in the following.

7×10^{31} seconds = 2.2×10^{24} years
The half-life of ^{128}Te

The element tellurium, which has 52 protons in its nucleus, has an isotope that can undergo double beta decay, which happens extremely slowly. It decays into the noble gas xenon, which has two protons more in its nucleus. The element in between (with 53 protons), iodine, does not have an isotope into which tellurium could decay. Tellurium was discovered in 1782, and looks grey and metallic. Both tellurium and selenium have chemical characteristics that makes it akin to sulfur, because the outer electron layer is almost complete, save for two electrons.

Tellurium crystals

Chapter 21

The Dark Eternities
10^{32} seconds: to infinity and beyond

There is not much to say about extremely large timescales — after 10^{32} seconds, more than 10^{24} (a septillion) years. Galaxies will have broken apart or collapsed almost completely into black holes. Most of them will be so far away, in the outer limits of the universe, that they are far beyond the horizon of what we can detect, and we will never see or hear from them again. Any remaining planets or stars will have cooled off completely and their temperatures will be approaching absolute zero. Even so, there are much larger timescales about which physicists have something interesting to share. These we discuss briefly in this chapter.

10^{41} seconds = 3.17×10^{33} years
2×10^{41} seconds
The minimal lifespan of the proton

The proton is a subatomic particle present in the nucleus of every atom, and thus ubiquitous on Earth and in our universe. Ernest Rutherford discovered the proton in 1919, though it is named after William Prout, who predicted its existence in 1815: the name is also inspired by the Greek word *protos* meaning 'first'.

It has a positive electrical charge and a mass of 1.67×10^{-27} kilograms; the diameter is about 1.65×10^{-15} meters. It is often accompanied by its partner in the atomic nucleus, the neutron, which closely resembles the proton, but is electrically neutral. In isolation, the neutron is not stable, as we explained in Chapter 3.

Ernest Rutherford (1871–1935)

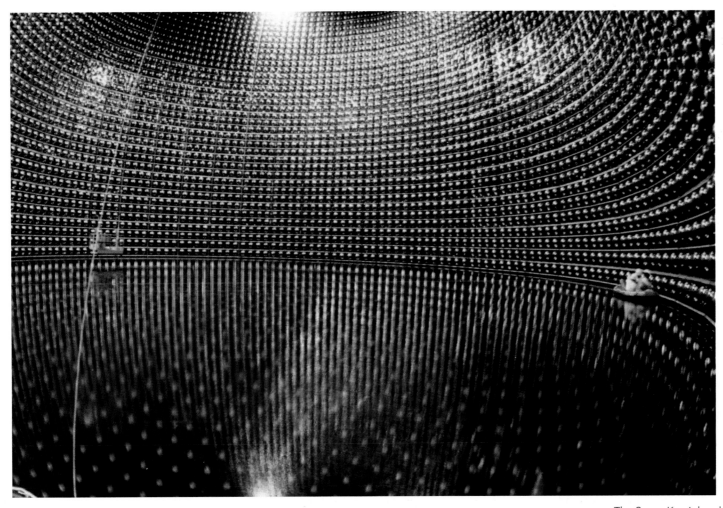

Scientists continue to try to determine whether protons are stable or not. The decay of a proton has not yet been observed experimentally, which has led to the lower limit for its half-life to be set as 6.6×10^{33} years. The lifespan of a free proton might be much longer, but according to the so-called Grand Unification Theories, an eventual decay of protons is inevitable. This means that every atomic nucleus would, eventually, be unstable.

The stability of a proton is of crucial importance for the existence of stable atoms, and therefore for the existence of all matter surrounding us — including the atoms in our own bodies. In fact, the far, far future of our universe depends to a large extent on the lifespan of a proton. It is therefore of interest to continue with experiments that attempt to determine its stability. Using the most sophisticated research, such as the Super-Kamiokande experiment in Japan, proton decay has still not been observed, and this led to the minimum value of its half-life to be set at what was said above, 6.6×10^{33} years. As yet there are

The Super-Kamiokande in Japan. A giant tank containing 50,000 liters of pure water was built in a zinc mine in Japan, about a thousand meters underground. Detectors are built into the walls to measure signals caused by proton decay.

no direct indications to suggest that the proton is unstable. Theoretically, in today's version of the Standard Model of elementary particles, protons do not decay. (Strangely enough, though, the Standard Model does allow for protons to decay *three at a time*, leaving behind pions, positrons and neutrinos. However, this would be a 'tunneling' effect that would be much rarer than the expected single proton decay).

Protons, just like neutrons, are not point particles, but are composed of so-called quarks. A proton consists of three quarks: two up quarks and one down quark, which are bound together by a type of particle that functions like glue: the aptly-named gluon. Quarks always exist in groups and cannot be isolated from one another. We will resume our narrative on quarks in Chapters 23 and 24.

In the so-called Grand Unification Theories (GUTs), which unite three of the four fundamental laws of nature — the electromagnetic force, the weak and strong nuclear forces — the proton is not stable. It is believed that the three forces are indistinguishable in the early universe, and that they are unified in a single primordial force, not including gravity (see Chapter 22). The various fundamental building blocks, quarks and leptons (such as the electron) are also no longer distinguishable from each other, meaning they should be able to transmute into one another. The GUTs predict the existence of a new particle (referred to in the picture on the top right as the X-particle),

which, through its interactions, causes the proton to disintegrate into a pion (consisting of a down quark and its antiparticle) and a positron (the antiparticle of an electron). Other decay modes might also be possible.

The half-life of a proton in a GUT depends on the X-particle's mass and other parameters of the model. Generally speaking, it is believed that the lifespan of a proton would not exceed the experimentally-determined lower limit by much, but this is far from certain. Future experiments, such as the planned Hyper Kamiokande, may determine whether the proton does in fact decay as the GUTs predicted.

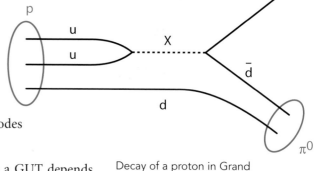

Decay of a proton in Grand Unification Theories.

10^{58} seconds = 3.17×10^{50} years
The lifespan of a black hole with a mass similar to that of Earth

A black hole is an object that is created when a heavy star implodes under the influence of its own gravitational forces. Relatively light objects, such as the Earth, cannot implode into a black hole because their gravitational forces are not strong enough. In physics, it is conceivable that a black hole with a similar mass to the Earth could exist, but we would not understand how it had formed.

Many characteristics of black holes follow laws of nature that we understand very well, meaning we are able to determine details of these characteristics quite precisely. For example, we know that black holes emit matter particles. This phenomenon has been derived on purely theoretical grounds by the well-known British physicist Stephen Hawking. The first emitted particles have very little energy, which means they can only be photons or gravitons, the quantum particles of gravity. By emitting particles, a black hole shrinks, but this is a very slow process. A black hole with a mass similar to that of the Earth would have a diameter of about 18 millimeters, a little smaller than an American cent or a two Eurocent coin. Such a black hole would emit particles, but the loss of energy is so minute that the total mass would take only 10^{50} years to evaporate. We will take a closer look at the generation and emission of particles in black holes in subsequent sections.

3×10^{72} seconds $\approx 10^{65}$ years
As a result of quantum melting processes, all solid objects lose their shape and become round droplets

Solid matter keeps its shape. Particularly when temperatures drop to absolute zero all matter will be as hard as a rock — with the exception of liquid helium. Nonetheless, there is movement: according to quantum mechanics, all atoms and molecules in matter, even at a temperature of zero degree, continue to have residual energy, so-called zero-point energy. This enables atoms to change places, as long as you have enough time on your hands to see it happen. If you were to tape this process and play it back on fast-forward, it would appear that matter is melting. This is true for all types of matter, and so in the end, everything in the universe would eventually melt and

morph into spherical droplets or evaporate. This is likely to be a moot point though, because protons and neutrons, which make up all atomic nuclei, would probably have disintegrated long before then.

10^{74} seconds $\approx 3 \times 10^{66}$ years
Lifespan of a black hole of a few solar masses

Black holes are remnants of heavy stars at the end of their lives. When a star of about three solar masses is burned out, its nucleus will implode, creating a black hole. Matter density and gravity are then so high that nothing can escape. This 'hole' in space can be described by Einstein's general relativity theory. It states that the space in which stars and planets move is curved, and the curvature is proportional to the mass of the object. For a normal celestial object, such as a planet or star, this looks like the image on the right.

Deep inside a black hole, the curvature is so extreme that we speak of a hole in space time. Everything that falls into it is lost forever. Even light rays are trapped in this curved space and will not ever be able to find their way out, making it look as if the hole is black. According to the theory of general relativity, black holes cannot emit anything, they can only absorb.

Some black holes are much larger than just three or five solar masses. It is presumed that most galaxies have a black hole at their center with dimensions corresponding to millions of solar masses. Everything that comes into the vicinity of such a black hole, including planets, stars or even clusters of stars, will be sucked in mercilessly. The absorption feeds the black hole, causing it to become even more ferocious

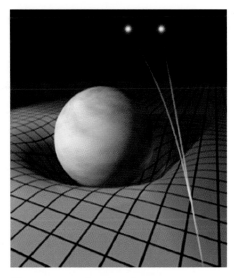

The presence of matter changes and curves its surrounding space. Approaching objects must follow a path in that curved space, which is effectively like experiencing attraction. A beam of light does not follow a straight line, but is curved by gravitational forces, in a way foreseen by Einstein and later verified by measurements.

Curvature of space caused by a black hole. A particle or a ray of light that approaches a black hole too closely — beyond the so-called horizon — will not be able to find its way back. Its only future is to fall further into the hole, where no one will ever see it again. Deep within the black hole the curvature is infinite; we are not able to describe what happens there, but because no one will ever be able to observe the regions surrounding this point from the outside, it does not really matter.

and unrelenting. Luckily, our solar system is far away from the center of our galaxy, so we will be safe for the foreseeable future. With a few simplified calculations we are able to derive that a black hole would devour a galaxy such as our Milky Way in a time span of roughly 10^{30} years. By then, all its stars will have fizzled out anyway and some will even have become black holes themselves.

The story of black holes is not yet finished. Einstein's theory of general relativity does not take quantum mechanics into account, which is necessary for a detailed description of certain characteristics of black holes and quantum gravity in general. When Stephen Hawking tried to give a quantum mechanical description of black holes, he made a remarkable discovery: black holes cannot be 'black' entirely. By quantum mechanical effects, particles are spontaneously created close to the horizon, and these are able to escape from the black hole's gravity. He calculated that black holes do radiate particles (now called Hawking radiation) and this radiation has a temperature that can be precisely calculated.

The temperature of a black hole is extremely low and depends entirely on its mass. The relationship is inversely proportional: the larger the mass, the lower the temperature. For a black hole of a few solar masses, the temperature lies in the area of 100 nK (a nanokelvin is 10^{-9}K). In comparison to the cosmic background radiation of the universe, which is 2.7K, this is so immeasurably low that, initially, a black hole will absorb more cosmic background radiation than it will emit.

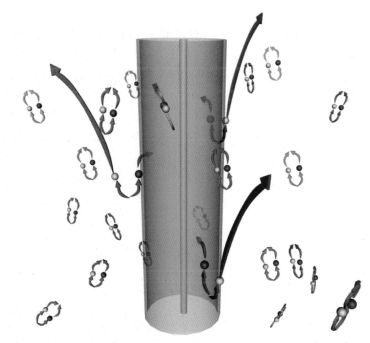

Eventually, however, the universe will cool down as well, and there will be a moment in time when the cosmic background radiation is lower than the Hawking temperature. From that moment, a black hole will lose matter, and it is relatively easy to calculate how long it will be before a black hole has completely evaporated: for a black hole of a few solar masses, this takes approximately 10^{66} years.

We must emphasize here that our knowledge of black holes is incomplete. Hawking radiation calculations still use certain simplifications and assumptions that are difficult to justify without a theory including quantum gravity. It may be that so-called String Theory (see also Chapter 22) — which attempts to unify quantum mechanics with relativity theory — will provide further clarity on these matters in the future. Only if we have a reliable theory for quantum gravity will we be able to unravel the more obscure mysteries of black holes.

According to the principles of quantum mechanics, fluctuations can arise in empty space where particles and antiparticles are created. Normally, they annihilate each other quickly, but if this process takes place close to the horizon (the grey area of the figure above) of a black hole, the center of which is indicated by the red line in the figure, where the time axis runs vertically, it is possible that a particle is sucked into a black hole, but not its antiparticle. The latter can escape the black hole and is radiated away. The energy for the creation of this particle/antiparticle pair is extracted from a black hole, which makes it look as if the black hole is radiating and losing energy. In time, a black hole can evaporate completely because all its energy (and thus its mass) is converted into Hawking radiation in this manner.

10^{90} seconds ≈ 10^{83} years and much longer timescales

Poincaré-recurrence time

The universe will have practically ceased to exist. Nonetheless, even longer timescales play a role in formal calculations of a statistical nature. A simple example is the following. Imagine particles moving around in a closed region of space. The number of possible positions the particles can occupy is enormous, even if the number of particles is not that large. It is then fairly simple to calculate how long it takes before those particles are in the exact same position again. Assuming they will take up pretty much all other positions before returning to their original one, it is a question of counting. A few examples are mentioned in the timeline on page 101. For a simple drop of water, which contains quadrillions and quadrillions of particles, the time it takes for them to assume the same position relative to one other, would have to be expressed in a septillion decimals: we would not be able to write such a large number in full in this entire book.

A surprising conclusion

The timescales of these last sections are so unimaginably long that only the most modern physical theories may yield any insight into them. Remarkably, these specialty areas of physics, which deal with such enormously long time spans, also touch upon extremely short time frames, such as proton decay. Proton decay can be examined theoretically in the context of the unified field theories. It might well

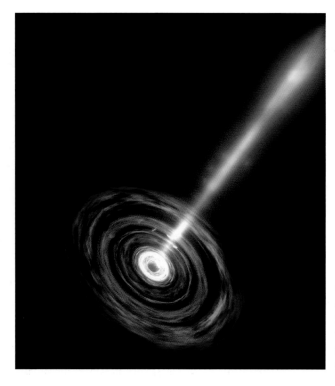

Gamma flash. The final destination of a super heavy star, an implosion resulting in a black hole. During the implosion, high energy gamma radiation is emitted in the direction of the rotational axis of the star.

take 10^{34} years before an average proton decays, but the processes responsible for its decay are extremely fast, much faster than most other interaction processes among particles.

Processes described by quantum gravity — such as the final evaporation of a black hole or events during the Big Bang — play out in extremely short timescales. Just like the 'dark eternities', these unimaginably quick events are shrouded in uncertainty. So much is still unclear, but this fascinating topic of ultrafast phenomena is heavily and enthusiastically speculated upon in scientific publications. In the next chapters we move to the extremely short timescales, starting with the 'Planck unit of time' of 5.4×10^{-44} seconds. We then move up by factors of 10, until we arrive back at the time span with which we started our book, that of one second.

All Timescales on a Timeline

All timescales on a timeline, from the shortest possible intervals (bottom left) to the longest ones (top right). The brown part concerns a period of which a lot is known, the violet parts are the extremely short intervals of which there is a lot of speculation, and the turquoise parts are the 'dark eternities', which cover a much longer period than even the current age of the universe. In the naming of the phenomena, the same color scheme is adhered to as in the rest of the book: **pink for decay times**, **blue for orbital times**, **green for timescales describing the evolution of the universe**, **orange for light-seconds and light-years**, and finally **yellow for vibrations and periodical occurrences**. Scattered among the coloured entries are other phenomena with characteristic time spans that do not belong to a specific theme.

The timeline clearly shows the relationships between the various timescales: the ratio between the very shortest imaginable timescale (Planck time) and the shortest time span observable with the largest particle accelerator is just as large as the ratio between the particle accelerator and a picosecond; the distance between a picosecond and a theatre production is again just as long. Jump with another such factor (approximately 10^{15}), and we have a period longer than one hundred times the age of the universe. The dark eternities, discussed only briefly in this book, comprise much larger steps in time (these are taken exponentially on the timeline).

As an aside, lots of contradictions in predictions exist in the timescales of the dark eternities. The melting of matter by quantum vibrations, for example, can only occur if protons do *not* decay. Further, the Earth shall only be absorbed by the Sun through gravitational radiation if the planetary system was not disrupted by other neighboring stars long before that. Therefore, the assertions are not all reliable as predictions of the future; they are really only illustrations of fairly abstract calculations. The actual future is unknown to the authors.

Time (seconds)

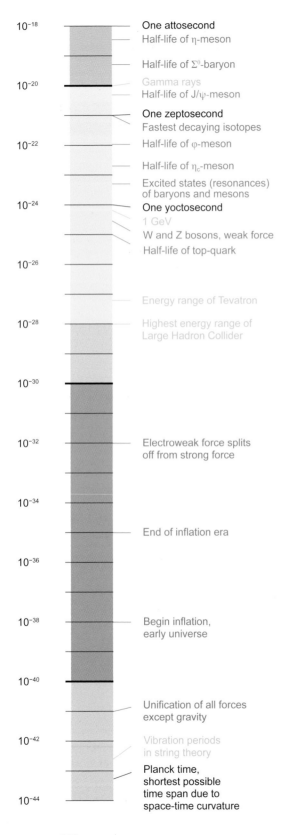

10^{-18}	One attosecond
	Half-life of η-meson
	Half-life of Σ^0-baryon
10^{-20}	Gamma rays
	Half-life of J/ψ-meson
	One zeptosecond
	Fastest decaying isotopes
10^{-22}	Half-life of φ-meson
	Half-life of η_c-meson
	Excited states (resonances) of baryons and mesons
10^{-24}	One yoctosecond
	1 GeV
	W and Z bosons, weak force
	Half-life of top-quark
10^{-26}	
	Energy range of Tevatron
10^{-28}	Highest energy range of Large Hadron Collider
10^{-30}	
10^{-32}	Electroweak force splits off from strong force
10^{-34}	End of inflation era
10^{-36}	
10^{-38}	Begin inflation, early universe
10^{-40}	Unification of all forces except gravity
10^{-42}	Vibration periods in string theory
10^{-44}	Planck time, shortest possible time span due to space-time curvature

Time (seconds)

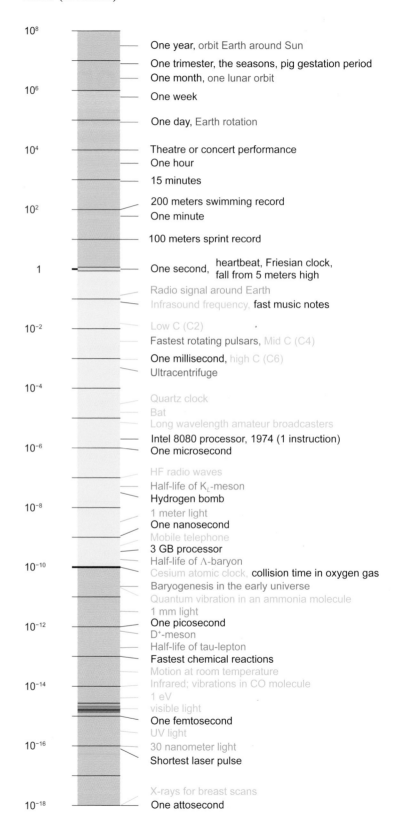

10^8	One year, orbit Earth around Sun
	One trimester, the seasons, pig gestation period
	One month, one lunar orbit
10^6	One week
	One day, Earth rotation
10^4	Theatre or concert performance
	One hour
	15 minutes
	200 meters swimming record
10^2	One minute
	100 meters sprint record
1	One second, heartbeat, Friesian clock, fall from 5 meters high
	Radio signal around Earth
	Infrasound frequency, fast music notes
10^{-2}	Low C (C2)
	Fastest rotating pulsars, Mid C (C4)
	One millisecond, high C (C6)
	Ultracentrifuge
10^{-4}	Quartz clock
	Bat
	Long wavelength amateur broadcasters
	Intel 8080 processor, 1974 (1 instruction)
10^{-6}	One microsecond
	HF radio waves
	Half-life of K_L-meson
	Hydrogen bomb
10^{-8}	1 meter light
	One nanosecond
	Mobile telephone
	3 GB processor
	Half-life of Λ-baryon
10^{-10}	Cesium atomic clock, collision time in oxygen gas
	Baryogenesis in the early universe
	Quantum vibration in an ammonia molecule
	1 mm light
10^{-12}	One picosecond
	D+-meson
	Half-life of tau-lepton
	Fastest chemical reactions
	Motion at room temperature
10^{-14}	Infrared; vibrations in CO molecule
	1 eV
	visible light
	One femtosecond
	UV light
10^{-16}	30 nanometer light
	Shortest laser pulse
	X-rays for breast scans
10^{-18}	One attosecond

Time (seconds)

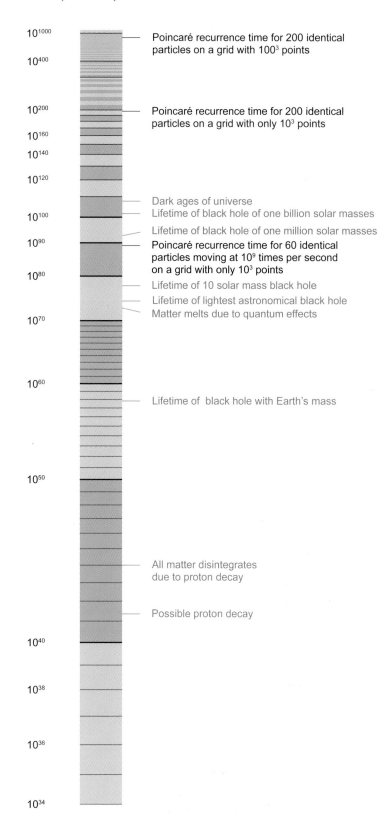

10^{34}

10^{32} — Half-life of tellurium-128

One septillion years

Earth devoured by Sun
due to gravitational radiation

10^{30}

One sextillion years

10^{28}

Double beta decay of selenium-82

Galaxies disintegrate

10^{26}

One quintillion years

10^{24}

Planetary system disturbed
by the stars

One quadrillion years

10^{22}

Last stars

Half-life of osmium-184

Lifetime of lightest red dwarf star

10^{20}

One trillion years

Local group becomes
one single galaxy

10^{18} Half-life of lutetium-176

Age of our universe

Origin of Sun, Earth and Moon

One billion years

Orbital period of Sun in Milky Way

10^{16}

Continental drift, dinosaurs

Distance to Virgo cluster, first horse

First elephant

10^{14} Pleistocene

One million years, *Homo erectus*

Homo sapiens

10^{12}

Precession of Earth's axis

Oldest trees, Dolmens

Great Wall of China

10^{10} Giant tortoise Adwaita

One century

Orbital period of Saturn

10^{8} Olympiad

Time (seconds)

10^{1000} Poincaré recurrence time for 200 identical
particles on a grid with 100^3 points

10^{400}

10^{200} Poincaré recurrence time for 200 identical
particles on a grid with only 10^3 points

10^{160}

10^{140}

10^{120}

Dark ages of universe

10^{100} Lifetime of black hole of one billion solar masses

Lifetime of black hole of one million solar masses

10^{90} Poincaré recurrence time for 60 identical
particles moving at 10^9 times per second
on a grid with only 10^3 points

10^{80} Lifetime of 10 solar mass black hole

Lifetime of lightest astronomical black hole

Matter melts due to quantum effects

10^{70}

10^{60} Lifetime of black hole with Earth's mass

10^{50}

All matter disintegrates
due to proton decay

Possible proton decay

10^{40}

10^{38}

10^{36}

10^{34}

Chapter 22

Small Timescales
10^{-44} to 10^{-26} seconds

From the longest timescales in the previous chapter we now jump to the very shortest, in which processes that take place go much faster than the blink of an eye, much speedier than lightning, or much quicker than the fastest computers. However, the fundamental physical ideas behind both extremely long and short timespans are linked, and some can only be studied and described by the most modern concepts in present day research. These theories — for example, quantum gravity — are nowhere near completion, and that is why the discussion of our next subjects is rife with speculation.

Moreover, the smallest timescales we are able to actually measure last about 10^{-16} seconds, which is a long way removed from the minute periods of around 10^{-44} seconds that we are about to get into. But the universe works to a schedule that is virtually incomprehensible: it can squeeze a whole history into the shortest time we can imagine. And there are a number of other ultra-fast phenomena of which we are gaining an increasingly detailed understanding.

Max Karl Ernst Ludwig Planck received the Nobel Prize in 1918 for physics.

10^{-44} seconds = 0.000,000,000,000,000, 000,000,000,000,000,000,000,000, 01 seconds

The shortest conceivable timescale for any physical process is putatively the Planck time. The German physicist Max Planck (1858–1947) noticed that there are three fundamental constants, which play an essential role in nature:

— Planck's constant, $\hbar = 1.0546 \times 10^{-34}$ kg m^2/s

— Newton's constant of gravity,
  $G = 6.672 \times 10^{-11}$ m^3/kg s^2

— and the speed of light,
  $c = 2.997,924,58 \times 10^8$ m/s.

One combination of these constants has the dimension of length, the Planck length, another corresponds to the Planck mass, and the third combination has the dimension of time:

$$\sqrt{\frac{\hbar G}{c^5}} = 5.39 \times 10^{-44} \text{ seconds.}$$

Because both the gravitational constant as well as the Planck constant have a very tiny value, while the speed of light is very large, the equation shows that this unit of time must be extremely small. It is generally assumed that this Planck time is the smallest possible interval of time. This might well be the shortest amount of time that nature requires to effectuate any change whatsoever. This sort of interval won't be measurable any time soon. The smallest time frame we are currently able to measure lies somewhere in the vicinity of 100 attoseconds, or 10^{-16} seconds. More about the attosecond in Chapter 30.

In a Planck time, light travels a distance of 1.62×10^{-33} centimeters, which is called the Planck length. At this scale a complicated structure of different physical processes emerges, as quantum phenomena (that is the dynamics of atoms, molecules and subatomic particles) are no longer negligible; gravity, which causes curvature of space and time, becomes the dominant force, and the speed of light, which determines how this curvature propagates, both play important roles. At other timescales, these three factors are not all simultaneously interlinked.

While a precise formulation of the laws of physics at this scale is not yet possible, many scientists believe that we should look in the direction of superstrings.

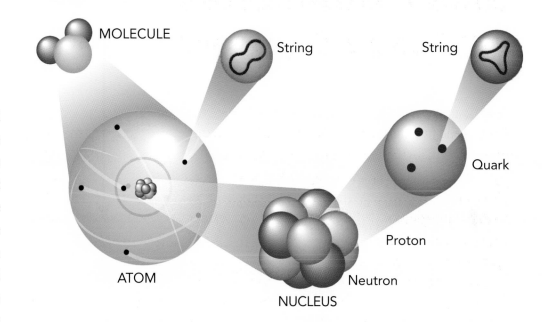

String theory assumes that elementary particles, such as electrons and quarks, are represented as vibrating strings. The various particles resemble strings with varying frequencies. The strings are believed to be as short as the Planck length, 1.62×10^{-33} centimeters, or perhaps just a little longer, and they vibrate at around the Planck frequency, or once every 5.39×10^{-44} seconds. A remarkable characteristic of string theory is that it appears to combine gravity with quantum mechanics and to provide a framework within which elementary particles and their mutual forces are unified. Many theoretical physicists hope that string theory will provide a precise and comprehensive description of nature at the Planck scale. With current experiments we are 'only' able to investigate distances of 10^{-18} centimeters; therefore strings are not directly observable. Consequently, it will be quite some time before we are able to test string theory experimentally.

These are string-like structures that create a nine-dimensional space and a one-dimensional time. Of the nine spatial dimensions, six get rolled up into a complicated knot, and the remaining three form the space as we know it. Vibrations in these strings take the form of elementary particles, which only assume a tangible form after a considerable time. The prefix 'super' refers to the fact that a special type of symmetry — 'supersymmetry' — is required to be able to describe these quantized strings in a consistent manner. More about supersymmetry in the section on 10^{-28} seconds.

10^{-43} seconds = 0.000,000,000,000,000, 000,000,000,000,000,000,000, 000,000,1 seconds

The universe must have been created in an extremely short time. The Big Bang theory was first coined in 1931 by Belgian priest Georges Lemaître (1894–1966). According to this theory, everything we observe in our universe today must have been compressed into a tiny ball of about 10^{-33} centimeters (the Planck length), which then expanded. The temperature within this Planck ball must have been extremely high, in the order of 10^{32} kelvins, the Planck temperature. Scientists have deduced that the ball must have been this small from the fact that all regions of the universe appear to be exactly the same age — if the age of the oldest stars and galaxies in all directions and regions is the same, then they must have spread from a single, central point and, therefore, have a common origin and past.

The laws of quantum mechanics appear to preclude a precise definition of the exact moment the Big Bang took place. This is why some scientists suspect that the very first moment in time is not a single 'point-event' in 'space time', but a vaguely defined area in our space-time continuum. We still know very little about this period, because there is as yet no accepted

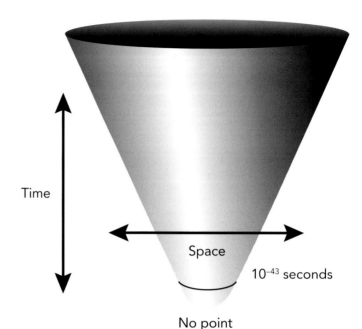

Schematic depiction of the way in which time and space may have been created as a result of the Big Bang.

theory of quantum gravity. It may even be that the concepts of time and space lose their meaning at the Planck scale; they are 'co-created' with the Big Bang, as suggested by the depiction above.

Perhaps there was no true 'beginning' at all; maybe the universe was 'relaunched' after an even earlier chapter in its history. These are, as yet, wild speculations. Furthermore, we do not know whether the universe, including the parts we will never be able to observe, is infinite or finite, as described in Chapter 20.

The period between the moment of the Big Bang — from zero until 10^{-43} seconds — is called the Planck era. It is a period of time about which we know next to nothing. The density of matter and the temperature at the Big Bang are infinitely high. The curvature of space is infinite, and none of the laws of physics known to us are valid at this point. Perhaps in the future, string theory will be able to elucidate the Planck era.

Moving forward in time, we are able to describe the universe's expansion in some detail using Einstein's relativity theory: we have stumbled onto more familiar ground. The descriptions that follow in the next few sections are the result of attempts to extrapolate all that we currently know and believe, and apply them to these timescales. Little is known about the smallest timescales so, here, we will have to move with relatively large steps.

10^{-38} to 10^{-32} seconds after the Big Bang

After the Big Bang, the expansion of the universe must have occurred at enormous speed. At this point in time, the universe is still smaller than the tiniest particles known to us now. According to most models, the universe is expanding exponentially until about 10^{-35} seconds after the Big Bang. This is called the period of inflation. The arguments in favor of inflation are very convincing: all parts of the universe, however far removed, appear to have a common past (see the section of 10^{-36} seconds after the Big Bang). The universe as we know it today is so big, and the Planck length so small, that there must have been a period of exponential growth. The forces responsible for this growth are attributed to the matter having a density many times higher than anything we are currently able to reproduce in laboratories. The effects of this show many resemblances to 'dark energy', which is presently responsible for the accelerating expansion of the universe (see also Chapter 20, about the future of the universe).

10^{-36} seconds after the Big Bang

From our present knowledge, this is about as long as the 'inflation' of the universe would have lasted: its period of exponential growth. After inflation, the universe was suddenly a few centimeters in size. We call this the 'horizon' of the universe. Within the horizon, the universe is quite homogenous and the temperature is the same everywhere. Without inflation theory it would have been a great mystery as to how this can be, because light

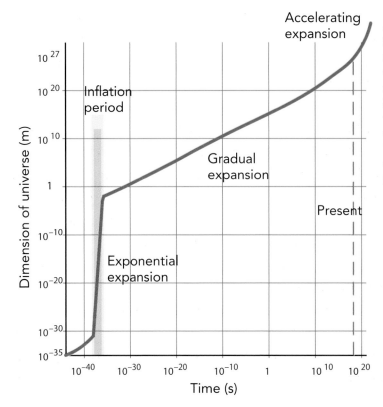

A simplified rendition of the idea of inflation. In extremely short timescales the universe expands exponentially; its radius grows with more than 30 powers of 10. Thereafter the inflation halts and the expansion slows down. Recent data shows that the expansion of the universe is now accelerating again.

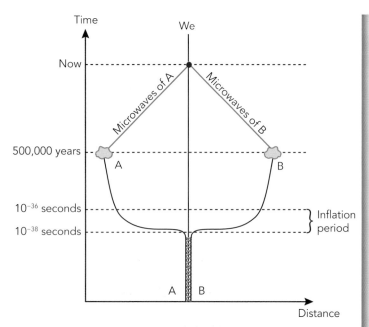

Two sources, A and B, appear to be too far removed from each other to have been in contact causally. However, observations indicate that the two sources behave similarly and have the same characteristics. According to the theory of inflation this is explained by the fact that information could have been shared at the very early stages of the universe.

in the short period of 10^{-43} seconds (the Planck era) could not have travelled more than 3×10^{-27} meters, which is much less than the size of the horizon. No information can go faster than the speed of light. What caused all those stars and galaxies to be formed at the same moment? We must assume that all distances started out very tiny. Then, while the universe expanded, the light waves went with it. This may explain why the universe became so homogeneous, while all causal contacts between the different regions got lost. They still

have a common past and are therefore strongly correlated (see figure on the left).

Next, it is possible to calculate how small fluctuations in space expanded along with inflation. Later, these became the seeds for the formation of galaxies and galaxy clusters, and also led to differences in the intensity of background radiation. The latter can still be detected in our universe at present. By measuring these variations extremely precisely, we furnish ourselves with valuable data about the inflation period and it is likely that, while you are reading this, the idea of inflation is confirmed beyond speculation. The precise moments of the beginning and end of inflation are still very uncertain. Perhaps inflation ceased as early as 10^{-32} seconds after the Big Bang.

Until 10^{-38} seconds after the Big Bang, each particle in the universe had more than 10^{16} GeV (giga-electron volts, see chart on the next page) of energy, and the temperature at that moment was about 10^{29} kelvins. Calculations show us that with this tremendous amount of energy per particle, all subatomic forces — except for gravity — must have had the same strength, and this is why they call this period the Grand Unification Era (see also Chapter 21, regarding proton decay).

After inflation, a period commences of which we know little, because the energy per particle is still so high that we cannot really perform any measurements. The inflation period must have ended abruptly, which resulted in a phase transition. Latent energy in the universe was transformed into heat, which would have caused the temperature to spike briefly, up to about 4×10^{28} kelvins. Density of matter must have been extremely high — about 10^{76} kilograms per cubic centimeter — and can probably be best described as a 'hot soup', in which quarks and gluons (the force particles of strong interactions) are able to move freely and have not yet been encapsulated in protons. The characteristics of this 'quark–gluon plasma' are investigated at the Relativistic Heavy Ion Collider (RHIC) on Long Island, New York, and in the Large Hadron Collider (LHC) in Geneva, albeit at lower densities and temperatures than in the early universe (10^{12} K at RHIC and somewhat higher at LHC). More about the 'quark–gluon plasma' can be found in Chapter 43. The universe continued to expand after inflation, but at a lower speed.

Gravity acts between particles with mass, and the strong force binds protons and neutrons together in an atomic nucleus. The weak force is responsible for radioactive decay, and the electromagnetic force works between particles with an electric or magnetic charge. At low energies these forces manifest themselves very differently, but it is suspected that the four stem from one fundamental force at extremely high energies. The unification of electromagnetic and weak forces occurs at energies of 100 GeV. Conversely, the segregation of this so-called electroweak force happens relatively late, at 10^{-12} seconds after the Big Bang. Earlier, after 10^{-32} seconds, the symmetry between electroweak and strong nuclear force was broken. (The disparity between these two

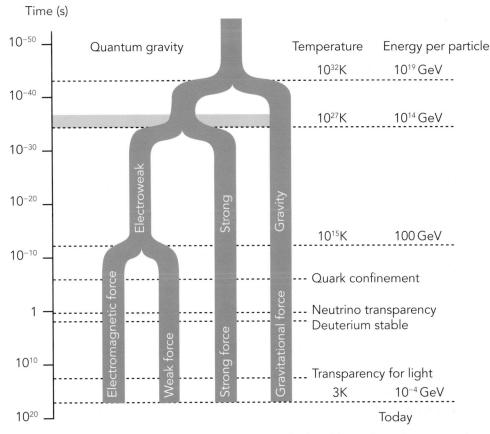

The four different forces in nature and their unification at high energies.

times and others has been deduced from the large differences in strengths of the forces currently being observed.) With the separation of the electromagnetic and weak forces, the period of the Grand Unification ends.

10^{-28} seconds

At the beginning of this chapter we discussed the possibility of using Planck's constant to convert timescales into energy scales, because time multiplied by energy always has the dimension of Planck's constant. The Planck time $T_{Pl} = 10^{-44}$ seconds

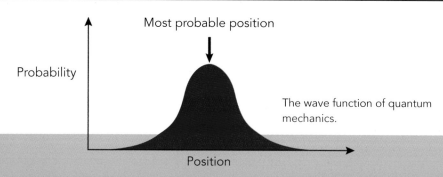

Most probable position

Probability

The wave function of quantum mechanics.

Position

PLANCK'S CONSTANT

The constant of Planck is central to quantum mechanics, the theory that describes the physics of atoms and elementary particles at microscopic level. At the end of the 19th century it became apparent that the heat radiated by an object at a given temperature could not be reconciled with the existing theories of thermodynamics.

Max Planck identified the solution to this problem: energy of radiation with a particular frequency is comprised of small packets, or *quanta*. This means that the energy of the radiation as a whole must be the sum of integer multiples of these elementary energy quanta. Later, with the aid of Einstein's work, it was understood that these energy quanta are actually 'light particles', and that electromagnetic radiation is therefore nothing more than a collection of light particles, or *photons*.

Planck deduced a formula from his experiments that determines the energy of a photon:

$$E = hf$$

where $h = 6.626 \times 10^{-34}$ kg m²/s = 4.1357×10^{-15} eV·s is Planck's constant, and f the frequency of the wave that describes the photon (physicists normally use the unit $\hbar = h/2\pi = 1.0546 \times 10^{-34}$ kg m²/s).

Subsequently, physicist Louis de Broglie discovered that Planck's formula is not only valid for photons but also for every other particle or object with a particular energy. Central to quantum mechanics is the idea that every wave can be described as one or more particles, but that every particle itself is also a wave, with a particular wavelength and frequency. The wave function determines the probability of identifying a particle at a certain location (see graph above).

The meaning of the formula devised by Planck and de Broglie is that a particle with energy E can be described by a wave with frequency f, and the relation is given by the formula above. Since the frequency of a wave expresses the number of oscillations per second, it can be computed how quickly the wave function of a particle with energy E vibrates. For instance, for a particle with an energy of 7 TeV — at this moment the highest energy at the LHC — the time for one oscillation is 6×10^{-28} seconds.

corresponds to the Planck energy $E_{Pl} = 10^{19}$ GeV (see also the Unification chart, on the previous page). Similarly, a timescale of 10^{-28} seconds corresponds to energies of 6.5 TeV (6.5 tera-electron volt = 6.5 trillion electron volt). Such energy scales are found at the LHC in Geneva.

In an underground tunnel, about 100 meters deep, two beams of protons are propelled in opposite directions and hurled against each other in circular tubes. After an acceleration to 99.999,999% of the speed of light, these protons reach a maximum energy of 7 TeV. The total collision energy of two particles is thus twice this amount: 14 TeV. From these collisions, scientists are trying to learn how fundamental physics at this scale works. The LHC experiment so far has focused primarily on discovering the Higgs boson, which explains how and why many matter particles obtain mass.

ELECTRON VOLT, ENERGY AND MASS IN ELEMENTARY PARTICLE PHYSICS

The electron volt (abbreviated as eV) is the amount of energy gained by the charge of a single electron (or any other particle with the same charge) moved across an electric potential difference of one volt. It is a much-used unit in elementary particle physics. A giga-electron volt (GeV) is 10^9 eV. The electron volt can be converted into the more familiar energy unit, the joule (J): 1 eV = $1.60217653 \times 10^{-19}$ J. The electron volt is a tiny amount of energy, but in particle physics it is often localized in an extremely small area. The energy produced in collisions between particles inside the LHC accelerator in Geneva reaches a maximum of 14 tera-electron volts, or 14×10^{12} eV = 1.4×10^4 GeV. The Planck energy is 10^{19} GeV. Mass and energy are related, via Einstein's well-known formula:

$$E = mc^2.$$

In this formula, E stands for energy, m for mass and c^2 is the square of the speed of light. Because of the extremely large value of c^2 in this formula, one gram of mass equals 8.988×10^{13} joules. That is the equivalent of the combustion energy of 15,000 barrels of crude oil. If we were able to convert a sugar cube of four grams into energy, then we would be able to solve all the Earth's energy problems with our sugar stock!

The mass of a proton is 1.672623×10^{-27} kilograms, or 938 MeV/c^2 (1 MeV = 10^6 eV).

The Planck energy, converted into mass, roughly equals 22 micrograms = 2.2×10^{-8} kilograms. If this mass was concentrated into a small area the size of the Planck length, 10^{-33} centimeters, it would form the smallest conceivable black hole.

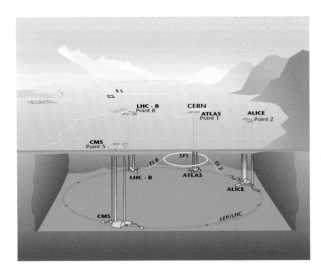

The underground tunnel where the LHC experiment takes place. Four large detectors have been placed where collisions between protons can be engineered. The names of the detectors are Atlas, Alice, CMS and LHC-B.

The first evidence of its existence was found on July 4, 2012 and confirmed on March 13, 2013. Theoretically, it was predicted in 1964, almost 50 years earlier. The existence of the Higgs particle is an important ingredient of the mechanism for breaking the electroweak force that unified the electromagnetic and weak forces in the early stages of the universe.

The LHC experiment searches for new particles that are not part of the Standard Model. For instance, searches are conducted for 'supersymmetry'. According to this theory, every particle in the Standard Model has a supersymmetrical partner. When the original particle is a boson (with an integer spin), then its 'superpartner' is a fermion (integer plus one-half spin), and vice versa. These

In reality, two beams of protons (not two single protons, as is often reported) are fired at each other. Each beam consists of many small bunches, and each bunch contains about 10^{11} protons (100 billion). The bunches are about 10 cm long with just a hair's thickness.

Under optimal circumstances, a maximum of about 2,800 bunches fits the rings, at a distance of about eight meters from each other. An enormous number of collisions occur each second (sometimes over 20 collisions in a single bunch crossing), with an enormous number of other particles being produced at each single collision (see figure on the next page). This is why extremely complicated statistical calculations are applied to the data that is generated. The large number of collisions is needed to produce enough material for a reliable statistical analysis of the measured data.

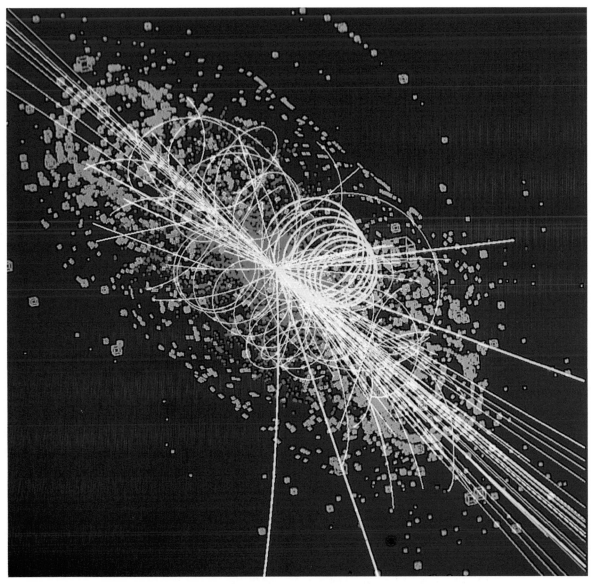

A computer simulation of the collision process of protons, where a Higgs particle is produced that subsequently disintegrates into all sorts of other elementary particles. Searching for the Higgs particle is like searching for a needle in a haystack because of the complicated collision processes. It was therefore a triumph for experimental particle physics when a new particle was discovered at CERN on 4 July 2012 and confirmed to be the Higgs particle in March 2013.

new particles would only be visible at energies of 10 TeV or more, which perhaps lies within the reaches of the LHC experiment. At the moment of writing, no signal supporting the existence of supersymmetry has been found. In fact, some theoretical models of supersymmetry have already been excluded by the LHC experiment. The discovery of supersymmetry would most certainly be considered an important scientific revolution within elementary particle physics.

Chapter 23

10⁻²⁵

10⁻²⁵ seconds

We have arrived in more familiar territory, where science has its foundations on firmer ground. At timescales of 10^{-25} seconds we find the half-lives of elementary particles that we know from the Standard Model. In the Standard Model of elementary particles, all matter consists of elementary building blocks, known as quarks and leptons. There are also additional particles that are responsible for the forces between quarks and leptons. Before discussing the decay times of top quarks and W and Z particles, let us look more closely at the Standard Model.

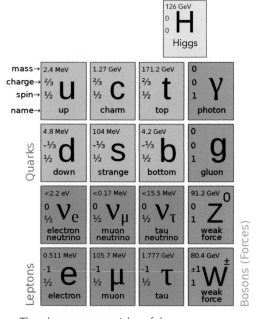

The elementary particles of the Standard Model. Every fermion also has an antiparticle, which is not shown in this figure. For example, the positron is the antiparticle of the electron, and the up antiquark is the antiparticle of the up quark.

There are six different quarks and six different leptons. Quarks always exist in groups and cannot be observed in isolation. Most matter on Earth is built from *up* and *down* quarks, which, when combined the right way, form protons and neutrons. These in turn may be combined to form atomic nuclei, around which electrons circle. The electron belongs to the family of leptons. Another lepton is the electron neutrino, which is related to beta decay. The other particles you can see in the table on the top right are more exotic and are usually only produced in particle accelerators or by forceful collisions of particles out in the cosmos.

Between leptons and quarks, forces are in play. Each interaction involves one particle that transfers its force to another particle. For example, a photon carries the electromagnetic force, while the Z and W bosons are responsible for the weak nuclear force. Gluons, on the other hand, are carriers of strong nuclear forces and bind quarks together in a proton or neutron.

What is missing from the Standard Model, is gravity. This is because no quantum-mechanical description exists yet for gravity. The quantum particle that should describe gravitational waves is called the graviton, but it cannot be detected experimentally. The interactions of the graviton are so weak that detection of individual gravitons in the foreseeable future is considered extremely unlikely.

The heaviest of all quarks is the top quark. Its mass — transformed into energy through Einstein's

$E = mc^2$ — is 172.6 ± 1.4 GeV/c². Very heavy and unstable particles decay easily into particles with a lower energy. This process can be observed in particle accelerators by shooting protons at their antiparticles (antiprotons), such as in the Tevatron accelerator of Fermilab, Chicago. On the right-hand side, such a process has been shown schematically.

10⁻²⁵ seconds
Roughly the half-life of a top quark

Top quarks and their antiparticles were first produced through proton-antiproton collisions. As shown on the right, top quarks decay via weak interactions into W bosons and bottom quarks. Top quarks are able to move freely for a very short time, between 10^{-24} and 10^{-25} seconds, before decaying. It is certain that the half-life must be shorter than 10^{-23} seconds — the timescale of strong interactions — because top quarks would otherwise 'hadronize' into groups, just like up and down quarks hadronize into protons and neutrons. Experiments show that this is not the case. Instead of hadronizing through strong interactions, the weak interaction acts faster and makes top quarks decay into W bosons and bottom quarks, which subsequently do hadronize.

When we say that quarks and gluons, formed at high energy collisions, 'hadronize', we mean that they collect other, secondary quarks and gluons to form clouds of hadrons (protons, pions, etc.). These clouds, called 'jets', spurt forth in the same direction and with the same total energy as the original quarks or gluons, which is how they are recognized in experiments.

2 × 10⁻²⁵ seconds
The half-lives of W and Z bosons

After top quarks, the heaviest elementary particles are the ones responsible for weak interactions — W and Z bosons (with the exception of the Higgs particle, whose mass is around 126 GeV/c²). Like many elementary particles, W and Z bosons manifest themselves in two ways. First, they can act as force-carrying particles, mediating the weak force between two particles, but they can also be created as actual particles themselves. They then have a mass of 80.4 and 91.2 GeV/c² respectively. The W particle is charged and thus there are two: the W⁺ boson with a positive charge, and its antiparticle, the W⁻ boson. The Z boson is electrically neutral. Both have an extremely short half-life: the W boson decays within 2.13 × 10⁻²⁵ seconds and the Z boson after 1.83 × 10⁻²⁵ seconds. Several different decay products are possible. Most of the time, such as depicted in the drawing above, the W boson decays into an electron and a neutrino, or a muon and its neutrino.

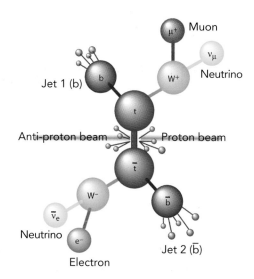

One of the possible ways a top quark and its antiparticle decay. In this experiment, a positively-charged muon, an electron and two jets are closely observed and registered.

Chapter 24

10⁻²⁴

10⁻²⁴ seconds = 1 yoctosecond

Yocto is an SI prefix† that is derived from the Greek *octo* (eight). It stands for 10^{-24}, or $1/1000^8$. It is the smallest officially-named prefix. The prefix yocto can also be used with other units. For example, a proton weighs 1.6 yoctograms. Along the same lines, a yoctosecond is the typical timescale encountered in the world of quarks and gluons. As mentioned in the previous chapter, there are six different quarks that have odd names with little meaning: up, down, strange, charm, bottom and top.

Quarks always huddle in groups because of the extremely strong interactions that bind them together. In other words, quarks are confined. There are many ways of forming little groups, usually a twosome (*mesons*) or threesome (*baryons*). None of the mesons are stable and the half-lives of a few of them are described in subsequent chapters. Mesons have integer spin and are therefore bosons. Baryons have a half-integer spin and are thus fermions. The best-known baryons are the proton and the neutron and their antiparticles, but the figure on the next page shows a few other possible combinations.

Quarks have a fractional electric charge, either ⅔ or −⅓, and come in three different colors. Of course, these are not 'colors' in the visual sense, but rather

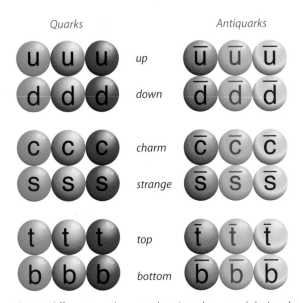

Quarks   *Antiquarks*

The six different quarks — and antiquarks — and their colors.

---

†SI stands for Système International (in French), the International System of Units. An SI prefix is metric, which precedes a basic unit of measurement to indicate a decadic multiple or fraction.

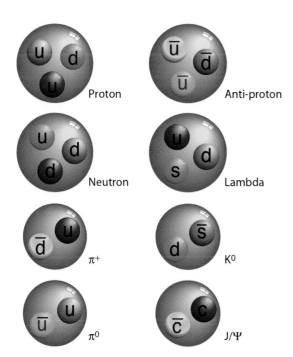

Baryons consist of three quarks, mesons of a quark and an antiquark.

has two up quarks and one down quark. The yellow colored 'springs' indicate the exchange of gluons, during which it is possible that pairs of quarks and antiquarks form and are subsequently annihilated. The arrows indicate the direction of their spin. The total spin of a proton is ½, resulting from a complicated distribution of spinning motion between quarks and gluons.

3.12×10^{-24} seconds
The half-life of the ρ-meson

One of the best-known mesons is the pion, of which there are really three: the electrically-neutral pion π⁰ and the positively- or negatively-charged pions, π±. As can be seen in the figure below, the π⁺ is comprised of an up quark and a down antiquark. The two quarks spin in opposite directions, such that the pion itself does not have a spin ($J = 0$). The rho, or the ρ-meson, has the same quark composition as the pion, but has a spin $J = 1$. Along the same lines, the ρ⁺-meson consists of an up quark and a down antiquark, but this time with an identical spin direction (see figure below).

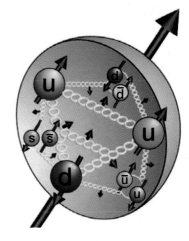

Cross-section of a proton.

an abstract term to help describe quarks' properties. However, as a whole, baryons and mesons (which are made up of quarks) are colorless. Antiquarks have the opposite charge and 'anti color', also described as 'conjugated colors'. The 'color language' is therefore convenient, since the three primary colors allow us to make colorless particles.

Forces between quarks are caused by the exchange of gluons, the force particles of strong interactions — a process that usually takes about one yoctosecond. It is quite turbulent within a baryon or meson; quarks are continually in interaction with each other by exchanging gluons that tie the lot together. On the top right we see a typical profile of a proton, which

π⁺
140 MeV

The charged ρ-meson and pion (π⁺).

ρ⁺
770 MeV

Because of the difference in spin, the ρ-meson is much heavier than the π-meson: 770 MeV/c^2 versus 140 MeV/c^2 (1 MeV = 10^6 eV, and we convert mass into energy via $E = mc^2$). It is remarkable that a difference in spin results in such a difference in mass, but this can be explained by the energy of the gluons in each of these configurations. We say that the ρ-meson is an 'excited state' of the pion, or a pion 'in resonance'. It therefore disintegrates very quickly into the pion: ρ$^+$ → π$^+$ + π^0, with a half-life of a bit more than three yoctoseconds.

3.87 × 10^{-24} seconds
The half-life of the Delta resonance

Another resonance — the first baryonic resonance to be discovered, now 60 years ago — is the Delta resonance. This is the first excited state of the proton (or the neutron), with a mass of 1,232 MeV/c^2. This is why the particle is also sometimes denoted as Δ(1232), using the Greek letter for Delta. The proton and neutron masses amount to 938 MeV/c^2 (the neutron is a little heavier). In 99% of the cases the Delta resonance decays into a nucleon (a proton or a neutron) and a pion, and in only 1% of all cases it decays into a nucleon and a photon. Remarkable about Δ is its spin, which is $J = 3/2$. It is the lightest known particle with that spin. Just like a proton, Δ contains three quarks, but here they all spin in the same direction.

Another well-known resonance of the nucleon is N(1440), with a mass of 1,440 MeV. It decays after only 1.5 yoctoseconds. Resonances with a higher mass are generally speaking more unstable and decay faster.

10^{-24} seconds
Light travels 0.3 × 10^{-15} meters

Gluons have no mass, thus according to special relativity theory they move at the speed of light. In one yoctosecond, light travels a distance of 0.3 femtometers, or 0.3 × 10^{-15} meters. That is about the distance between two quarks within a proton, of which the diameter is 1.75 femtometers. The interaction process between two quarks occurs through the exchange of a gluon and the time that takes is in the order of one yoctosecond.

Chapter 25

10^{-23}

10^{-23} seconds = 10 yoctoseconds

In the next three chapters we will predominantly encounter the decay processes of elementary particles or light isotopes. These are the timescales during which a number of the most unstable mesons decay. The pion, with its mass of about 135 MeV/c², the lightest meson, has a wave function with a vibration time of about 30 yoctoseconds. This is also the time span within which many strong interaction processes in atomic nuclei occur.

10 yoctoseconds = 3 femtometers = 3×10^{-15} meters
The distance light travels in 10 yoctoseconds

Atomic nuclei are typically this size. In fact, all these numbers are related: if the pion had been lighter, atomic nuclei would have been larger, because the protons would tend to get separated further. The diameter of the smallest atomic nucleus — that of hydrogen, with only a single proton — has been established quite precisely and is 1.75 femtometers. In 10 yoctoseconds, light travels therefore about the distance between two protons within an atomic nucleus. The heaviest nuclei, such as uranium, have diameters in the area of 15 femtometers. The radius of an atomic nucleus increases with the mass number (which is the total of the number of protons and neutrons) to the ⅓ power.

1.8×10^{-23} seconds
The half-life of the η_c-meson

The charmed-eta-meson, η_c, is a bound state of a charm quark with its antiparticle, the charm antiquark. The spin of the meson is $J = 0$, which

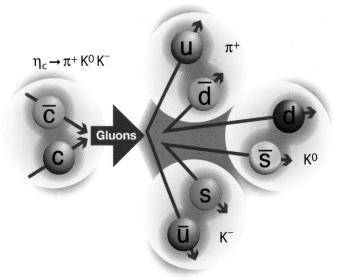

An example of one of the decay modes of the η_c-meson.

Mesons $q\bar{q}$					
Symbol	Name	Quark content	Electric charge	Mass GeV/c²	Spin
π^+	Pion	$u\bar{d}$	+1	0.140	0
K^-	Kaon	$\bar{s}u$	−1	0.494	0
ρ^+	Rho	$u\bar{d}$	+1	0.770	1
B^0	B-zero	$d\bar{b}$	0	5.279	0
η_c	Eta-c	$c\bar{c}$	+1	2.980	0

Information about some of the most common mesons and their quark compositions.

means the two quarks turn in opposite directions. Its mass is 2,980 MeV/c². It is a variant of the ordinary η- and η' -meson, which consist of combinations of quark-antiquark pairs of up, down and strange quarks. A bottom-eta-meson exists as well: η_b consists of a bottom quark-antiquark pair. This meson is much heavier, 9,389 MeV/c². Its decay time is unknown. Maybe now you would expect there to be a top-eta-meson, consisting of a top-top antiquark pair? In fact, this bound state does not exist. The reason for this was mentioned earlier in Chapter 22: the top quark decays within 10^{-25} seconds, and therefore it does not exist long enough to hadronize into a bound state.

Chapter 26

10^{-22}

10^{-22} seconds = 100 yoctoseconds

There are quite a number of mesons and baryons. Many elementary particles are symbolized by or named after a letter of the Greek or Latin alphabet. In the table on p.118 we show the most well-known mesons, which consist of up, down and strange quarks and their antiparticles. There are similar figures with mesons that have one or more charm, bottom and top quarks. In the 1960s, more and more new baryons and mesons were discovered, along with the system that became known as quark theory, or 'Quantum Chromodynamics' (QCD): baryons consist of three quarks, mesons of a quark and an antiquark, and antibaryons of three antiquarks. During the 1970s, the insight that quarks are held together by gluons was developed. Fluctuations create a background, known in scientific literature as a 'sea', of quark-antiquark pairs, which also assert themselves from time to time.

1.07 × 10^{-22} seconds
The half-life of the φ-meson

The phi meson, φ, consists of a strange quark-antiquark pair. Both quark and antiquark rotate around a parallel axis in the same direction. Their total spin is thus $J = 1$, and so we can find φ in the middle of the bottom figure on the right. Its mass is 1,019 MeV/c^2, a little bit heavier than

a proton, which consists of three quarks with a total mass of 938 MeV/c^2. Mesons, however, cannot decay into baryons, because their quark composition does not allow it. Instead, the φ-meson decays into K$^+$ and K$^-$ mesons (or kaons) about half the time, but it can also turn into other products. During the decay, the electric charge as well as the strangeness number is maintained. The strangeness number counts the number of strange quarks and strange antiquarks in mesons or baryons (see figure on the right).The half-life of a φ-meson is about 100 yoctoseconds. This is a relatively long time for a meson that falls apart as a result of strong interactions. We now know this is because the strange quarks are positioned quite close together because of their mass; the smaller the distance, the weaker the activity of the strong force.

The production of kaons in the decay process is relevant for the study of asymmetry in the universe between particles and antiparticles. We elaborate on this in Chapter 40.

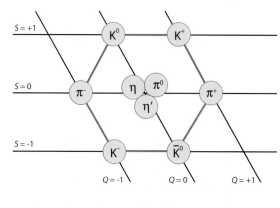

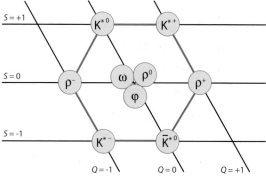

In the top illustration, mesons are shown with spin $J = 0$. These are formed by two quarks that spin around parallel axes in opposite directions. In the bottom illustration, we see mesons with spin $J = 1$.

The electric charge is denoted by the letter 'Q' and the 'strangeness' number by 'S'. The latter indicates how many strange quarks and antiquarks are present in the meson: S = −1 for the strange quark, S = 1 for its antiparticle, and S = 0 for all other quarks. The K-meson ('K' stands for kaon) consists of a strange quark and up antiquark with opposite spin, thus S = −1 and J = 0. We will discuss the various mesons in several parts of this book.

Chapter 27

10^{-21}

10^{-21} seconds = 1 zeptosecond

In this chapter, we make the transition from the yocto to the zepto eras. In 'zepto' we recognize the Latin term *septem*, which means 'seven'. Just as with the yoctosecond, the name comes from powers of a thousand. The zeptosecond is $1/1,000^7$ seconds. Its opposite is the zettasecond, which is 1000^7 seconds. The opposite of the yoctosecond is the yottasecond, 1000^8 seconds. As with the previous period, the most important natural phenomena that occur are the decay processes of mesons, but here we see the decay of extremely unstable isotopes as well.

1 zeptosecond = 300 femtometers =
3×10^{-13} meters
The distance light travels in a zeptosecond

In one zeptosecond, light travels a distance of 300 femtometers. A photon, or a light particle that is fired from the center of a hydrogen atom, leaves the atomic nucleus and moves towards the inner electron layer. In some ways, this distance appears to be quite large, because there is a vast expanse of empty space between an atomic nucleus and the electron closest to its center. Indeed, the distance to the closest electron is 5.3×10^{-11} meters. Compare this to the diameter of a proton, 1.32×10^{-15} meters. After one zeptosecond, light has only traveled about 1/100th of the distance to electron layers.

2.25×10^{-21} seconds
The half-life of the η'-meson

We have already encountered the family of η-mesons in Chapter 25. In the illustration in Chapter 26 we see two η-mesons with spin $J = 0$ and strangeness number $S = 0$: the η-meson and the η'-meson (eta-prime-meson). The η'-meson is almost twice as heavy, with 958 MeV/c^2 compared to 548 MeV/c^2 for the η-meson. This is surprising because both mesons contain the same sort of quarks: a mixture of up, down and strange quarks, and their antiparticles. It took quite a while before this difference in mass was explained by the theory of quarks and gluons, the theory of quantum chromodynamics. Because of its higher mass, η' has a shorter half-life than η — a little more than two zeptoseconds.

The quickest decaying isotopes: helium-5, hydrogen-7 and lithium-4

In the zepto era we also encounter the most unstable isotopes. Helium-5 has a half-life that has been determined quite precisely: 0.76×10^{-21} seconds. It consists of three neutrons and two protons (see illustration on the right).

There exist other exotic, but extremely unstable isotopes. For example, scientists were able recently to produce a very heavy version of hydrogen, ^{7}H: an atomic nucleus with one proton and six neutrons. The decay time has not yet been exactly determined, but lies somewhere in the region of one zeptosecond, maybe even a little faster. A similar decay time is attributed to the isotope lithium-4. In stable form, lithium has three protons and three or four neutrons. With a single neutron, ^{4}Li has an extremely short lifespan: a half-life of about one-tenth of a zeptosecond.

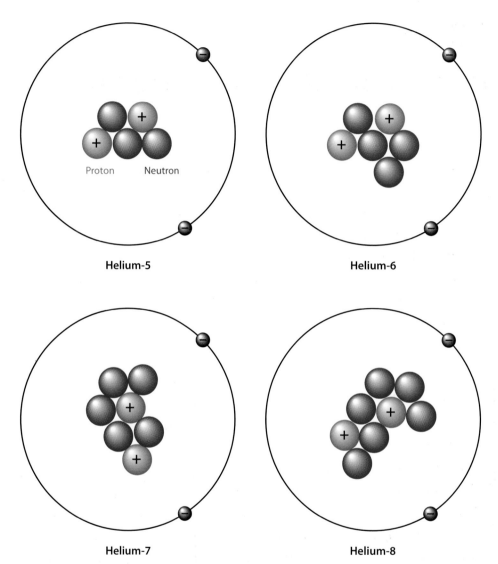

Helium-5　　　**Helium-6**

Helium-7　　　**Helium-8**

There are eight helium isotopes, of which only ^{3}He and ^{4}He are stable. Helium-4 is most common on Earth, and is produced as an alpha particle — ^{4}He atomic nucleus without electrons — from radioactive decay. An additional neutron creates the extremely unstable ^{5}He atom.

Chapter 28

10⁻²⁰

10⁻²⁰ seconds = 10 zeptoseconds

This is the first time we encounter frequencies of electromagnetic radiation. It forms a recurring theme in this book, because frequencies vary from extremely high (far beyond ultraviolet) to extraordinarily low (from infrared light to the longest frequencies of radio signals). Frequencies correlate with timescales, as light and radio waves have a period of one divided by the frequency. The timescale of 10 zeptoseconds corresponds to the wave period of the most energetic gamma rays — more about this at the conclusion of this chapter.

0.70×10^{-20} seconds
The half-life of the J/ψ-meson

The sub-nuclear particle that is known under the remarkable name J/ψ or J/psi-meson is a composite particle, comprised of a charm quark and its antiparticle. There are multiple ways in which these two quarks can be bound in a state called 'charmonium'. This is analogous with the 'positronium', the bound state of an electron and its antiparticle, the positron.

The lightest charmonium particle is the charmed-eta-meson (η_c, see Chapter 25). This particle is less stable than the J/ψ particle because η_c easily decays into two gluons. While J/ψ is heavier (meaning it contains more energy, and would normally be less stable), it is actually more

stable because it needs three gluons or a photon to decay. J/ψ is easier to observe experimentally because, at the right collision energy, these particles are produced copiously when an electron, e⁻, collides against a positron, e⁺. After the collision, J/ψ quickly decays back into an electron and a positron. J/ψ was the first observed charmonium particle (in 1974) and with it came the first compelling proof of the existence of the charm quark.

0.81×10^{-20} seconds
Wave period of light created as a result of electron-positron annihilation

The positron is the antiparticle of the electron. Both have a mass that represents an energy of $E = mc^2 = 511$ keV (kilo-electron volt). When a

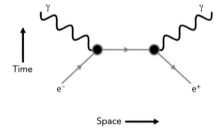

Feynman diagram of electron-positron annihilation. The freed light particles are gamma photons.

positron nears an electron, they may annihilate each other. Often, two gamma photons or particles of light are produced. The law of momentum conservation dictates that both photons shoot off in opposite directions, with the same amount of energy, meaning both will have exactly 511 keV. Using the relationship $E = hf$ it follows that they have a frequency of 1.235×10^{20} Hz.

1.21891 × 10⁻²⁰ seconds
The half-life of the Y-meson

The number of possible bound states of quarks is considerable. The various different mesons and baryons that can be created constitute a wealth of different elementary particles with large variations in characteristics and half-lives. The *Upsilon* or Y-meson consists of a bottom quark and its antiparticle. Discovered by the team of Leon Lederman in 1977, this was the first particle that contained a bottom quark. That same team had announced the discovery of a new particle a year earlier, but it proved not to exist. Afterwards, this particle was given the name 'Oops-Leon'. A year later, though, the team did strike gold.

Gamma radiation
Wave periods shorter than 10⁻¹⁹ seconds

Gamma rays are electromagnetic rays with frequencies above 10^{19} hertz (Hz, vibrations per second). They are created when an atomic nucleus undergoes nuclear reactions. Just as electrons circling the atomic nucleus are able to emit or absorb energy — creating photons with frequencies commonly much lower than 10^{19} Hz — nucleons (protons and neutrons) in an atomic nucleus are also able to emit or absorb energy. Because the atomic nucleus is much smaller than the atom itself, and its particles move much faster than the electrons surrounding the nucleus, the amount of energy is generally higher, roughly a million times higher. Accordingly, gamma rays have a much shorter wavelength — less than ten picometers or 10^{-11} meters — and a higher frequency than those other photons, but the underlying principles as to how they work remain the same.

Gamma rays rush to meet us from the depths of the universe. A spectacular phenomenon is the gamma ray burst. This occurs when super heavy stars explode, when two neutron stars fuse or when a black hole is formed. This is always paired with extremely energy-rich nuclear reactions between fast-moving nuclei. During a very short timescale, varying from milliseconds to minutes, high-energy gamma radiation is produced in small bundles that burst through the universe. These bundles are strongly focused because the particles are usually moving quickly along the magnetic lines of the source. More about this in Chapter 45.

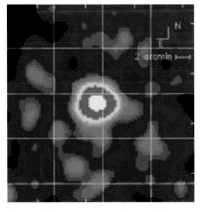

X-ray of the gamma ray burst of February 28th, 1997, captured by the satellite, BeppoSAX.

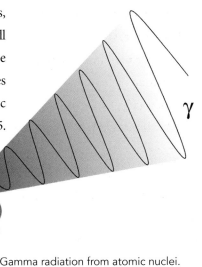

Gamma radiation from atomic nuclei.

Chapter 29

10^{-19}

10^{-19} seconds = 100 zeptoseconds

We are slowly but surely reaching timescales that we are able to conceptualize. A good example is X-rays. Wave periods of X-rays can be as short as 10^{-19} seconds. The corresponding wavelength is then 0.3 Ångström (Å), or 0.3×10^{-10} meters, and that is about the size of small atoms.

0.513×10^{-19} seconds
The half-life of the Σ^0-baryon

The Sigma-0-baryon, Σ^0 — with a mass of 1,192.64 MeV/c^2 (that is 27% heavier than a proton) — is closely related to Σ^+ and Σ^- (see Chapter 38). Nevertheless, its half-life is very much shorter, at only 0.513×10^{-19} seconds. This is because it is unstable as a result of electromagnetic disturbances, by emitting a photon it changes into the lighter lambda — or Λ — baryon, which has the same quark composition (see also Chapter 38 on 10^{-10} seconds).

Hard X-ray radiation, with wave periods of 0.3 to 3.3×10^{-19} seconds

The 'harder' variant of X-ray radiation has such a short wavelength that this light can be used to determine the position of atoms in crystals and molecules. Molecules are best crystalized first, as the use of X-rays is more effective this way: millions of identically arranged molecules then reflect the rays in the same manner. By analyzing the relative intensity ratios of the reflected light, it is possible to identify and recognize the position of the atoms in a molecule.

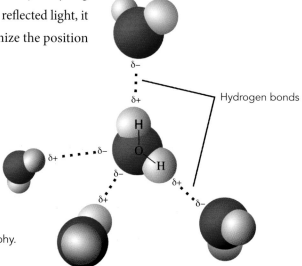

Three-dimensional image of the positioning of hydrogen atoms in water (molecules), as determined with the aid of X-ray crystallography.

X-rays are also referred to as Röntgen rays, after the German physicist who produced and detected electromagnetic radiation of this wavelength in 1895, an achievement that earned him the first Nobel Prize in Physics in 1901. Outside Germany and the Netherlands, the word X-ray, which was also used by Röntgen himself, is more common.

1.794×10^{-19} light-seconds
The Bohr radius

The most elementary and simple atom is that of hydrogen. It consists of a positively-charged proton in its nucleus and a negatively-charged electron that encircles it. Despite the attraction of the electrical force between the two charges, the electron is held at a distance from the proton.

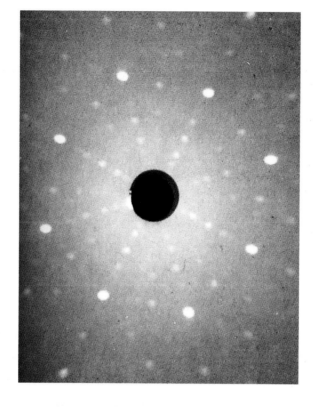

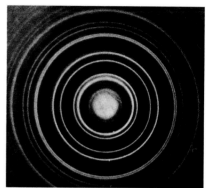

X-rays in crystalline materials are reflected at certain angles. This was discovered by the physicist Max von Laue, who received the Nobel Prize in 1914. This is because the wavelength of the radiation is comparable to the distance between atoms in a crystal. Consequently, the waves reflected by the atoms can only run in phase when they go in precisely defined directions, the directions of positive interference. By studying the patterns formed by the dots emerging at different angles, the arrangements of the atoms in a crystal can be revealed. When crystals in powder form are studied, the dots are smeared out into circles, the so-called Debye-Scherrer patterns (image on right).

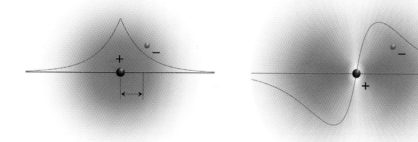

The hydrogen atom. According to modern quantum mechanics, the electron is not point-like but should be considered as a wave, a sort of spherical cloud surrounding the nucleus. On the left, the ground state, the S-wave ('S' stands for sharp here, because the wave function peaks sharply). The electron does not rotate around the nucleus (its orbital angular momentum is zero). The average distance of the electron is then the Bohr radius (indicated with an arrow in the left figure). On the right, the excited state: the P-wave ('P' stands for principle), whereby the electron rotates with an angular momentum of one around the nucleus. It is then not able to get close to the proton. The graphic on the left show how the wave function depends on distance of the electron to the nucleus.

According to the laws of quantum mechanics, the electron is only able to move in precisely defined energy orbits surrounding the proton. An electron has the lowest possible energy when it is in the innermost layer; we refer to that as the *ground state* of the electron. The average distance to the nucleus is the Bohr radius, 5.29×10^{-11} meters or 0.529 Ångström. A light signal takes 1.794×10^{-19} seconds to travel this distance.

4 to 8×10^{-19} seconds
Gamma radiation of the Mössbauer effect

The frequency of gamma rays emitted by an atom plays an important part in the so-called Mössbauer effect. The German physicist Rudolf Mössbauer investigated whether gamma photons emitted by an atomic nucleus could also be absorbed by the same species of atomic nuclei. Generally, this appeared not to occur: as a result of the large amount of energy in a gamma photon, both the emitting atomic nucleus and the target atomic nucleus experience a recoil, which causes energy to be lost. This is why the energy from the emitted gamma photon is insufficient to cause the other nucleus to get into the same excited state. This effect can be averted by using atoms that are fixed into a solid crystal lattice. The remaining atoms in a crystal block the recoil, as a result of which the expected resonance can occur. This has been represented schematically in the figure on the bottom right.

The remarkable aspect of this effect is that some atomic species take a relatively long time to emit and absorb such a photon. Therefore the frequency of the photon is fixed and determined very precisely, and the effect follows this frequency extremely accurately. Now, if either the source or the absorbing target is moving, the Doppler effect causes a shift in frequency, resulting in a change in absorption strength. If an atomic nucleus has multiple similar excited states, for example because it has a spin, then differing absorption lines occur in this manner (see the figure above). Energies of used photons lie at around 10 to 20 keV (kilo-electron volt). Using the Planck Einstein relationship — the energy of a photon is the frequency × Planck's constant — it becomes evident that the wave periods lie between 4 and 8×10^{-19} seconds.

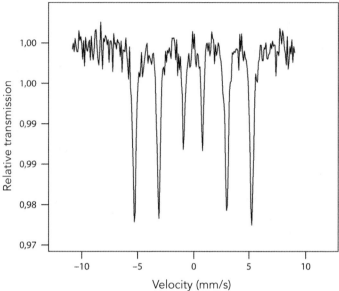

A typical pattern of the absorption lines of the Mössbauer effect.

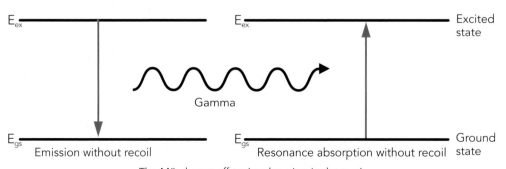

The Mössbauer effect (explanation in the text).

Chapter 30

10⁻¹⁸

10⁻¹⁸ seconds = 1 attosecond

The prefix 'atto' is derived from the Danish word *atten* for 18. In one attosecond, light travels a distance of about 0.3 nanometers = 3 Ångströms (1Å = 0.33 × 10⁻¹⁸ light-seconds), or about three hydrogen atoms. Photons of this wavelength have a frequency of 10^{18} Hz = 1 exahertz (1 EHz), and therefore an energy of about 4,136 electron volts — or about 4 keV — each.

To put these proportions into perspective: one attosecond is to a second, as a second is to twice the age of the universe.

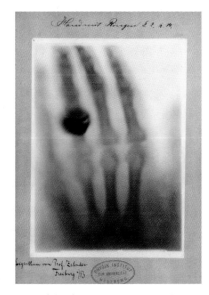

One of the first X-ray photographs. We see the outline of a hand belonging to Anna, Wilhelm Conrad Röntgen's wife, along with her wedding ring. The X-ray was made on December 22, 1895.

0.3 to 30 attoseconds
Wave period of 'soft' X-rays

Soft X-rays have wavelengths of between 0.1 and 10 nanometers, and therefore frequencies of $3{\times}10^{16}$ and $3{\times}10^{18}$ Hz. That is the area between ultraviolet (UV) and gamma rays. The soft X-ray variant is not very invasive — not even a micrometer in water. However, this radiation still ionizes and is therefore dangerous to our health.

The X-rays were so named by their discoverer, Wilhelm Conrad Röntgen (1845–1923), to signify an unknown type of radiation. In many languages, the radiation is actually named after him personally, 'Röntgen rays'. The terms Röntgen, X-rays and gamma rays are used interchangeably, but initially, the terms Röntgen and X-rays were used for radiation created by making electrons or ions jump over high voltages in a vacuum tube, and gamma rays when they came from radioactive nuclei. X-rays are created when electrons crash into a heavy metal target, such as tungsten or molybdenum. Radiation with a continuous spectrum is created when the electron comes close to an atomic nucleus, where it is deflected by its strong electric field, emitting a quantum ray.

We call that *bremsstrahlung*, German for 'brake-radiation', radiation that is created when an electron rapidly decelerates (negative acceleration). The radiation is the energy that is released because of this action. The electron may also hit another electron buried deep inside the atom, so that it flies off together with the projectile. The hole that is created is immediately filled by a nearby electron, which makes a large energy jump and emits radiation as a result. Because the size of the energy jump is determined very precisely, the radiation's spectrum has sharply defined lines — a so-called discrete spectrum. Those wavelengths usually lay around one Ångström (0.1 nanometers). In the case of mammograms carried out to detect breast cancer, softer rays are usually used, with wavelengths of around three Ångströms, of which the frequency is 10^{18} Hz.

0.346×10^{-18} seconds
The half-life of the η-meson

The eta particle, η, with a mass of 547.85 MeV/c², is remarkable because despite being able to decay via the strong force, it has a relatively long lifespan: its half-life is 0.346×10^{-18} seconds. This is because the strong force allows the particle to decay into three pions only, which means its mass surplus is only 138 MeV/c². This is relatively little

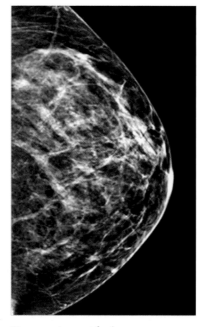
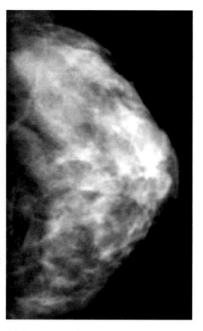

X-ray equipment for breast cancer research is constantly improving. *Left*: an X-ray obtained using current, digital techniques. *Right*: the results with older equipment.

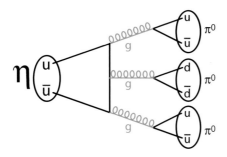

Feynman diagram for η-decay. The η-meson is actually a mixture: it fluctuates in being a uū pair, a dd̄ pair and an ss̄ pair. The π⁰ fluctuates between uū and dd̄. The forces affecting quarks are gluons (g), the particles that transfer the strong nucleus force according to the theory of quantum chromodynamics.

and therefore the decay process is relatively slow. The η-particle has several special symmetrical characteristics. It consists of the same quarks as pions, but is nonetheless a lot heavier. When this was first discovered, it appeared to clash with the theory of the strong subatomic force, quantum chromodynamics — QCD in short — but the deeper causes of this large mass difference were to a large extent identified at a later stage.

Chapter 31

10^{-17}

10^{-17} seconds = 10 attoseconds

Electromagnetic waves with a frequency of 10^{17} hertz have a wavelength of three nanometers. That also means that light travels three nanometers in 10 attoseconds. For a light wave that propagates with velocity c and frequency f, the wavelength λ obeys the following formula:

$$\lambda f = c.$$

The wave period belonging to this frequency is 10 attoseconds. The energy of these light particles according to Planck's relationship $E = hf$ is 413 electron volts, so that electrons exposed to this radiation are able to bridge more than 400 volts.

Ions and electrons move with higher velocities as temperatures rise. If the latter is higher than 10 million kelvins, then their energy comes into the area of kilo-electron volts (keV) and the radiation's wavelength will be a few nanometers. In the space around the Sun — as well as around more active stars much further away — areas with such high radiation do exist. In the vicinity of black holes, such as the one at the center of our Milky Way in particular, radiation of this spectral range appears to be emitted.

That said, this radiation does not penetrate the Earth's atmosphere very deeply. To enable observation of these areas in interstellar space, Röntgen detectors are launched in balloons, rockets or satellites. For example, on May 1st, 2008, NASA launched a rocket that is able to make Röntgen photographs in energy areas between 0.07 and 1 keV.

The Crab Nebula (see the picture on the next page) consists of matter emitted by an exploding star or supernova. By comparing varying wavelengths astronomers obtained a good picture of the physical processes that are evolving there. At the center of the nebula are the rapidly rotating remains of this exploding star, a pulsar (see Chapter 48).

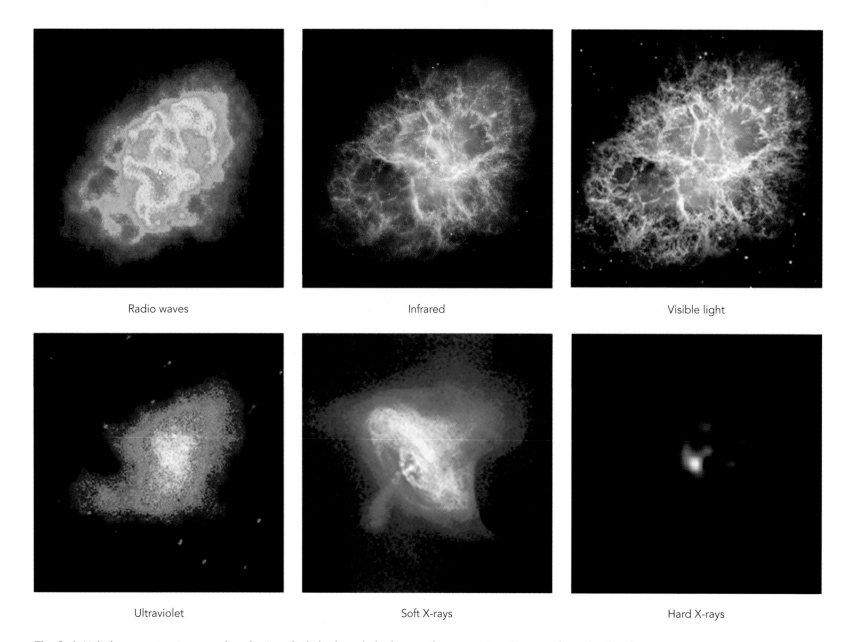

Radio waves

Infrared

Visible light

Ultraviolet

Soft X-rays

Hard X-rays

The Crab Nebula, seen at various wavelengths. In radio light the nebula shines with varying intensities — indicated with colors.
The remainder of the exploding star that formed the nebula is a pulsar, and the source is visible in the radio images (a small white dot).
In infrared and visible light we can clearly see how matter from the exploding star forms filaments. Red is radiation emitted by hydrogen,
blue radiation from extremely hot electrons. The center part of the nebula is particularly visible in ultraviolet or X-ray spectrum. This shows
us that the pulsar still heats up this area quite considerably.

Chapter 32

10⁻¹⁶

10⁻¹⁶ seconds = 100 attoseconds

▌ Light travels 30 nanometers in this time.

Electromagnetic radiation with this wavelength (namely 30 nanometers) vibrates with a period (P) of 10^{-16} seconds and is between 10 and 400 nanometers, better known as ultraviolet light (UV). Its wavelength is shorter than visible light (400–800 nanometers), but longer than X-rays. The frequency of UV light is therefore higher than the violet light that we can see with the naked eye, explaining the name.

According to the laws of quantum mechanics the energy of oscillating waves is always quantized. The size of these energy packets is inversely proportional to their oscillation period. The relevant formula is as follows,

$$E = h/P$$

where P is the period (the duration of one full oscillation), E the energy of the wave packet and h Planck's constant. A useful unit of energy is the amount that an electron gains as a result of passing through a voltage drop of 1 volt: 1 electron volt, or 1 eV, about 1.6×10^{-19} joules (see also the box at 10^{-28}

seconds). Planck's constant is 4.13567×10^{15} eVs. That means that at a frequency of 10^{16} Hz the corresponding energy is about 41 electron volts. The energy regime of wave packets of ultraviolet light is between 3 eV and 124 eV.

Many chemical reactions occur when an electron leaps across a few volts, so ultraviolet light generates many chemical processes — for example, in our skin when it is exposed to sunlight, which contains a lot of UV in addition to visible light. Too much ultraviolet light can be hazardous to our health.

A space telescope with the name Far Ultraviolet Spectroscopic Explorer (FUSE) was launched on June 24, 1999 by NASA as part of its *Origins* program. FUSE detected light with wavelengths of between 90.5 and 119.5 nanometers, which other telescopes are usually unable to observe. The primary goal of the missions was the detection of deuterium (heavy hydrogen), to identify how deuterium, which must have been created shortly after the Big Bang, was formed by stars. FUSE circles the Earth at about 760 kilometers.

There are many interesting phenomena that we can observe with UV light. We have included two examples in this chapter, photographed in visible light and in ultraviolet light.

The Andromeda Galaxy in visible light.

The Andromeda Galaxy in ultraviolet light.

The Andromeda Galaxy seen with infrared light.

The flower of a Geranium type, left as seen by the human eye, and right in ultraviolet light. This allows insects, which are able to see ultraviolet light, to find the heart of the flower.

67 attoseconds
The shortest laser pulses (2012)

In 2008, extremely short laser pulses were created that lasted 80 attoseconds (abbreviated 'as'), a period in which light is only just able to complete a full vibration or oscillation. To create such short pulses, we start with longer ones (in the order of a femtosecond). Then, multi-layered mirrors are placed in its path, which force the front of the pulse to take a longer route than the trailing part, because the front penetrates the material of the mirror more deeply than the tail. This has the effect of shortening the pulse. Short pulses created this way can match the movement of atoms and even electrons within an atom. Thus, the attosecond laser can be used as a camera or stroboscope, to capture the fast movements of atoms and electrons. With quickly-improving techniques there is little doubt that this new record will be broken again in forthcoming years.

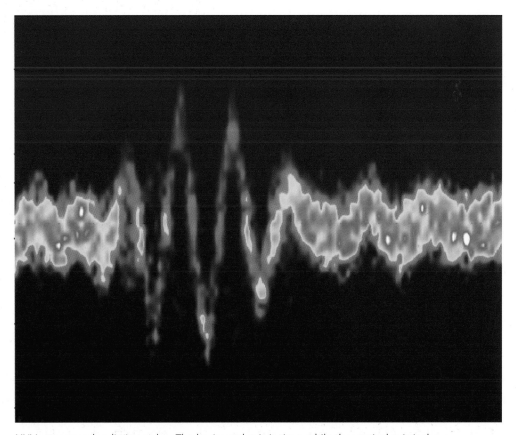

XUV attosecond radiation pulse. The horizontal axis is time, while the vertical axis is the intensity of the laser pulse. The distance between the two green peaks correlates with a laser cycle of about 80 attoseconds.

Chapter 33

10^{-15}

10^{-15} seconds = 1 femtosecond

The prefix 'femto' is derived from the Danish word *femten*, which means '15'.

▌ In one femtosecond, light travels a distance of 300 nanometers.

Light of this wavelength has an energy quantum of more than four electron volts. The frequency with which each vibration lasts for a femtosecond is 10^{15} hertz, or one petahertz (PHz). This is the frequency of ordinary, visible light.

One femtosecond

The response time of an electron in an atom

In atoms and molecules, electrons move with velocities of 1/1,000th to 1/100th the speed of light. That means they are able to move from one atom to the next in a femtosecond. Therefore, this is the maximum speed of a chemical reaction between two neighboring atoms.

▌ 2.329,764,111,719,67 × 10^{-15} seconds
The vibration time of a characteristic frequency of the strontium atom

The American Joint Institute for Lab Astrophysics in Boulder, Colorado, has been able to measure this vibration time extremely precisely by capturing a few thousand very cold strontium atoms in bundles of laser light. It was particularly important to isolate them from interfering magnetic fields.

The vibration time was measured so precisely that we would be able to redefine the second as the time that a strontium atom needs to complete 429,228,004,229,952 vibrations. Even though other atoms can be made to vibrate in very precise frequencies as well, the definition of the second as based on the cesium clock has not yet been revised (see Chapter 38).

▌ 1.26 to 2.63 × 10^{-15} seconds
The wavelength of visible light

Visible light has a wavelength of between 380 nanometers (violet) and 780 nanometers (red). The corresponding frequencies are 7.9×10^{14} Hz and 3.8×10^{14} Hz. Because the surface of the Sun has a temperature of about 5,700 kelvins, it emits light that peaks at about 600 nanometers, or 5×10^{14} Hz.

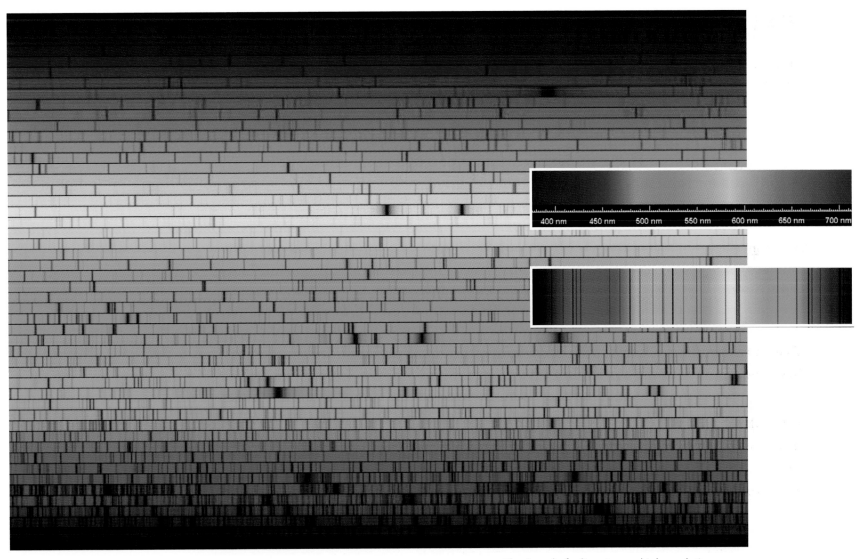

The Fraunhofer lines in very high resolution.
Inserts
Above: a continuous light spectrum of 380 to 720 nanometers. *Below:* the same spectrum, but with Fraunhofer lines that indicate which wavelengths are absorbed by the solar atmosphere.

That is right in the middle of the visible area. The Sun emits all wavelengths of visible light, after which certain frequencies are strongly absorbed by the Sun's surrounding atmosphere. This means that the solar spectrum shows a number of sharp, dark lines: the Fraunhofer lines.

Chapter 34
10^{-14}
10^{-14} seconds = 10 femtoseconds

In ten femtoseconds, light travels a distance of 0.003 millimeters, or three micrometers.

A micrometer (µm) used to be called a micron (µ). Before that an even older term was used, a millimillimeter (1/1,000th of 1/1,000th of a meter). These are the smallest distances visible (but only barely) with microscopes that use ordinary light. Important laser technologies have been developed in the femtosecond domain, such as the *frequency comb laser*. Its spectrum consists of a tremendously large number of extremely thin frequency lines at a set distance from each other, resembling a comb. This technique is used for precision measurements, for example in atomic clocks, but also to determine extremely small and short movements. John Hall and Theodor Hänsch received the Nobel Prize in Physics in 2005 for their contributions to the development of the optical frequency comb technique.

0.26 to 3.3 × 10^{-14} seconds
The wavelength of near infrared light

The frequency of so-called *near infrared light* (IR) lies between 3×10^{13} and 3.8×10^{14} hertz. Its wavelength then lies between 0.78 and 10 micrometers. It is no longer possible for humans to observe this with the naked eye. Heat radiation at a temperature of about 1,000 kelvins or 727 degrees Celsius is extremely strong and peaks at three micrometers or 10^{14} Hz, in other words at 100 THz, putting it

A rattle snake. Sensors are located between its nostril and eye, which are sensitive to infrared and heat radiation.

in the near infrared. Nonetheless, objects at this temperature appear 'red-hot', because a lot of visible light is emitted as well.

Infrared literally means less than red. It is light with a wavelength longer than that of red light, so our eyes cannot see it. But we can feel radiation emitted by warm objects, and many animals have organs that are able to detect this radiation much more accurately than we can. Because of the geometrical positioning of these sensors or organs, animals are usually able to determine quite precisely where the heat source is located; usually some desirable prey.

At almost all wavelengths, astronomers make observations in space. In Chapters 31 and 32, we show the results of research in the infrared domain, including the appearance of the Andromeda Galaxy in infrared light. The pictures below show how materials appear transparent through infrared light, but not with visible light; such as the common garbage bag. With other materials, for example a pair of spectacles, the reverse is true (see the same pictures, below).

1.55×10^{-14} seconds
The vibration time of the CO molecule

Molecules that consist of two atoms are kept together by a chemical binding force, created by a cloud of electrons. The atoms can exhibit two kinds of motion: vibration and rotation. The vibration frequency can be derived from the characteristic absorption lines that are exhibited by these materials when light shines through them. The vibration frequency is determined by the solidity of the chemical bonds between the two atoms. Carbon monoxide, CO, has a vibration period of just a little less than 10^{-13} seconds. Other bi-atomic molecules, for instance O_2, have comparable vibration frequencies. The rotation can also be identified from the light spectrum. Rotation velocities are quantized, according to the rules of quantum mechanics, and lie at around 10^{11} rotations per second. The fact that rotations are usually slower than vibrations can be understood from quantum mechanics: the energy needed to make a molecule rotate is considerably less than what is needed to make it vibrate, because the latter requires the chemical bond to be distorted. The formula $E = hf$ tells us that the frequencies of the rotations are accordingly slower than those of the vibrations. Only if you spin a molecule so fast that it gets distorted will its rotation frequencies match those of the vibrations.

At NASA, a man with spectacles and a garbage bag shows the difference between what is visible through normal and infrared light. Pay attention to the invisibility of the cool part of the garbage bag and the opacity of the spectacles' lenses in IR.

Chapter 35

10^{-13}

10^{-13} seconds = 100 femtoseconds

10^{-13} seconds = 0.000,000,000,000,1 seconds = 0.1 picoseconds

> In 100 femtoseconds, light travels a distance of 0.03 millimeters. This is less than half the width of a human hair.

Middle infrared light has wavelengths between 0.01 and 0.03 millimeters. Heat radiation with these wavelengths is emitted by objects at a very low temperature, around 100 kelvins, or −173 degrees Celsius. Radiation at this temperature has the strongest intensity at frequencies around 10 THz (terahertz).

2×10^{-13} seconds

The disintegration of iodine cyanide

Chemical reactions occur by forming or breaking chemical bonds between atoms. The time frame of these reactions varies from extremely short timescales that are barely observable, to long periods — think, for example, of a rusting nail. The fastest chemical reactions occur as quickly as the vibrations of atoms, roughly between 10 and 100 femtoseconds. With the aid of femtosecond laser spectroscopy we can analyze and image those vibrations and study the chemical reactions at microscopic level. This area of science is named femtochemistry. In 1999, the Egyptian-American scientist Ahmed Zewail received the Nobel Prize for his earlier work on one of the simplest chemical reactions, that of molecular disintegration. For this research he used the iodine cyanide molecule, which disintegrates into iodine (I) and cyanide (CN):

$$ICN \rightarrow I + CN.$$

By shooting at this molecule with a femtosecond laser pulse, disintegration is initiated. With a second laser pulse shortly after, it is possible to observe how the reaction has taken place and calculate how long it has taken. The response time for the disintegration of iodine cyanide proved to be 200 femtoseconds.

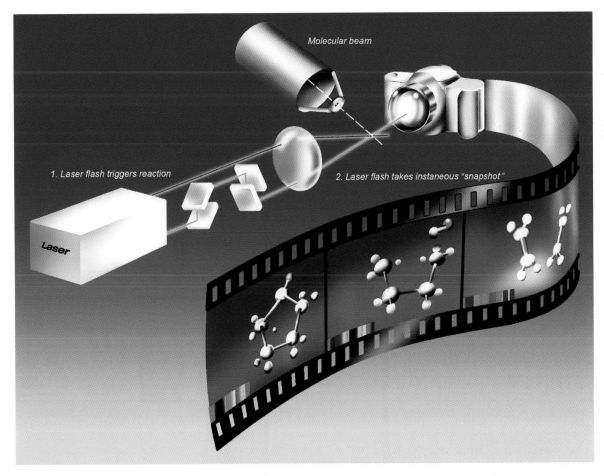

A schematic illustration of the femtochemics principle. With the aid of two femtosecond lasers, molecules' motion and chemical reactions can be depicted.

2.014 × 10⁻¹³ seconds
The half-life of the τ lepton

In addition to the muon (see Chapter 42), the electron has an even heavier brother, the τ lepton or tau lepton. It was discovered in 1975. A neutrino comes along with it, the τ neutrino, ν_τ. With a mass of 1,776.8 MeV/c², the τ lepton decays much faster than the muon: its half-life is 2.014×10^{-13} seconds. In 18% of all cases, this weak interaction process proceeds the same way as for a muon, but in 17.4% of all instances muons or

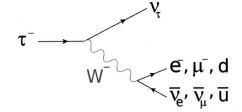

pions can be created. It is believed that these are the only leptonic relatives of the electron, because any unforeseen 'black sheep' would clash with the beautiful agreement of our current theories and experiments.

Feynman diagram of the tau-decay. Decay products may vary. A tau neutrino is always emitted, and then pairs with an electron with its anti neutrino, a muon with its anti neutrino, or one of several pions (consisting of quarks); sometimes a photon is released as well. The decay is caused by the weak force, via the negatively charged W particle.

At the conclusion of this chapter we provide an overview of the electromagnetic spectrum with a chart. We have dedicated a lot of attention in this book to the timescales and frequencies that belong to electromagnetic radiation. The timescales that play a role here reach a range of 15 powers of 10: from 10^{-4} to 10^{-19} seconds.

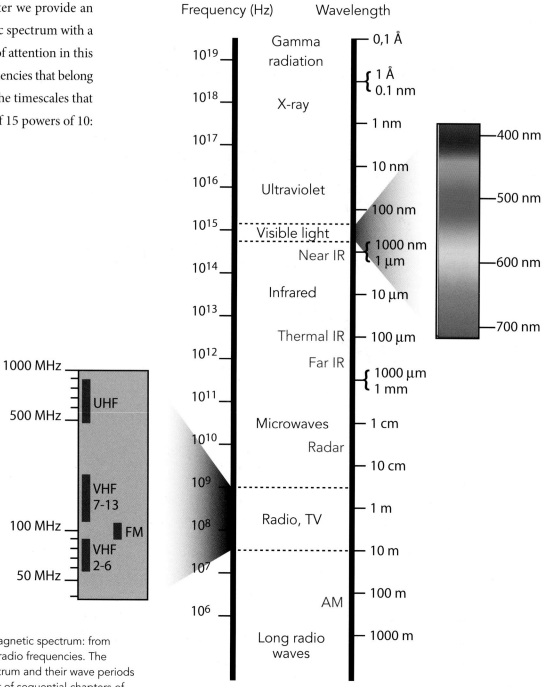

Overview of the electromagnetic spectrum: from gamma radiation to long radio frequencies. The specific areas of this spectrum and their wave periods are discussed in a number of sequential chapters of this book (see Chapters 28–46).

Chapter 36

10^{-12}

10^{-12} seconds = 1 picosecond

The picosecond, referring to the Italian *piccolo* (very small), is abbreviated to 1ps. It is 1/1,000,000,000,000th, or 1/1,000,000th of 1/1,000,000th of a second. Light travels a distance of 0.3 millimeters in this time, about the width of the dot at the end of this sentence. With picosecond laser pulses, researchers can document and photograph ultra-short movements of molecules like the oscillations in vibrating molecules in hydrogen bridges in water, which vibrate with a period of about one picosecond.

One picosecond
Wave period of the terahertz frequency

Far infrared electromagnetic radiation is part of the sub-millimeter domain. It is light with wavelengths of between 0.03 and 0.3 millimeters. At a temperature of about 10 degrees above absolute zero (10 kelvins), weak electromagnetic radiation is emitted that peaks at about 1 THz = 10^{12} hertz. The wavelength of this radiation then lies around 0.3 millimeters and its vibration period is one picosecond.

The fastest transistors work at frequencies in the vicinity of the terahertz area. These transistors are not yet available commercially. They only work at extremely low temperatures. A transistor is used to amplify an electric signal or as an electric switch that can be turned on or off. The terahertz transistor can switch in just one picosecond. The fastest transistors in current computers process signals with frequencies of three to four gigahertz (1 GHz = 10^9 hertz), see Chapter 38.

0.72×10^{-12} seconds
The half-life of the D^+ meson

1.135×10^{-12} seconds
The half-life of the B^- meson

A meson is a subatomic particle that contains a quark and an antiquark, and as such is related to pions and kaons. The D^+ meson has a mass of 1,869.6 MeV/c^2 and a half-life of 0.72×10^{-12} seconds. This subatomic particle contains a charm quark, c. The word charm is used in the sense of 'magic': there were problems with observations of how K particles decayed, but when researchers assumed that a mysterious *charm* quark existed, these observations made sense. Scientists confirmed the existence of charm quarks much later on, but the name stuck. The charm quark decays quickly, usually by forming a *strange* quark, and a K meson typically appears. Because in this case the mass surplus is huge, these types of decay processes occur faster than the usual weak interactions.

Mesons that contain the bottom quark, *b*, are even heavier: the B$^-$ meson has a mass of 5,279 MeV/c^2. Nonetheless these mesons have about the same, long half-lives: the B$^-$ meson, which consists of an up-antiquark and a bottom quark (B$^-$ = $\bar{u}b$), has a half-life of 1.135×10^{-12} seconds. This is because the *b* quark evidently does not transition easily into other quarks. All the baryons that contain a *c* quark or *b* quark have half-lives of about 10^{-12} seconds or a little bit shorter.

With the decay of the B$^-$ meson an interesting phenomenon occurs. In the Standard Model of elementary particles, the B$^-$ meson decays into a lepton and an antineutrino, via the intermediary, negatively charged W$^-$ boson. Often this is the τ lepton and the tau antineutrino $\bar{\nu}_\tau$ (see figure above), but other decay products are possible. Therefore, experimentalists are trying to measure the chances of each mode of decay, to understand in how many cases the different types of decay occur.

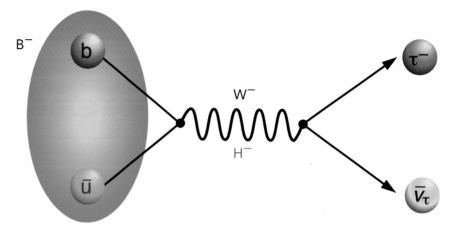

Diagram of one of the decay modes of the B$^-$ meson, B$^- \rightarrow \tau^- + \bar{\nu}_\tau$. The decay products are a tau-lepton and a tau antineutrino. In the Standard Model without supersymmetry, this decay can only occur via the weak force, by exchanging a negatively charged W boson. In the supersymmetric version of the Standard Model there is another possibility: the exchange of a negatively-charged Higgs particle.

In the theory of the Standard Model with supersymmetry, another type of decay is possible. The supersymmetric Standard Model has not one but five Higgs particles: the neutral H^0 that also exists in the Standard Model, two new neutral Higgs particles, one positive and one negatively charged Higgs, H$^\pm$. Via Higgs particle H$^-$ the B$^-$ meson can decay into a τ lepton and a tau antineutrino $\bar{\nu}_\tau$. This would mean that the supersymmetric Standard Model allows for more decay modes. Precisely measuring the decay fractions of the B$^-$ is therefore very important for the discovery of new physics, such as the phenomenon of supersymmetry (see the section hereafter, as well as Chapter 22 under 10^{-28} to 10^{-26} seconds).

1.4×10^{-12} seconds after the Big Bang

The universe is now about 100 million kilometers in size. The temperature is around 4×10^{15} kelvins and its density is 10^{28} kilograms per cubic centimeter. So far most particles are behaving similarly to each other, but now differences start to emerge. Up to this point in the universe's life, supersymmetry may have existed, a structure that dictates that particles with an integer and half-integer spin (bosons and fermions) are not distinguishable from each other. But this is not the case now, meaning that supersymmetry must have been disrupted somewhere. We refer to this as 'broken supersymmetry'. According to calculations, this breakdown must have occurred at or just before 1.4×10^{-12} seconds after the Big Bang — otherwise we would have identified this symmetry in our current experiments. The energy per particle must have been about one TeV (10^3 GeV), which corresponds to the energy that is released when a singly-charged particle such as the electron passes through a voltage difference of one trillion (10^{12}) volts.

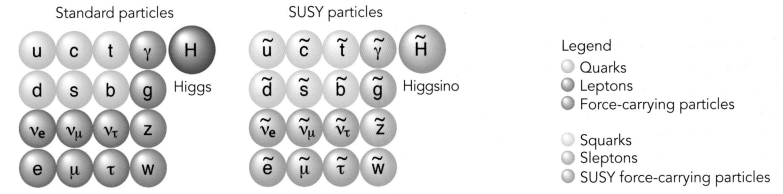

On the left are the familiar elementary particles from the Standard Model. On the right their supersymmetric partners (called SUSY particles). There are six quarks and six leptons (matter particles), all with half-integer spin (fermions). The supersymmetric partners are bosons with integer spin. They are called 'squarks' and 'sleptons'. The force-carrying particles are also doubled in the supersymmetric Standard Model. The photon, or light particle, has the 'photino' as its superpartner, and so forth. In a theory with supersymmetry, both partners have the same mass, while the spin differs by half a unit. If the concept of supersymmetry exists in nature, it must be broken at present and the superpartners must have become much heavier. They form the most likely candidates to explain dark matter in the universe. Verifying the existence of supersymmetry — or disproving it — is an important task of particle accelerators such as the Large Hadron Collider (LHC) at CERN, in Geneva.

Chapter 37

10⁻¹¹

10⁻¹¹ seconds = 10 picoseconds

In 0.000,000,000,01 seconds, or 10 picoseconds, light or a radio signal travels a distance of three millimeters. That is the distance between the two arrows, below:

The fastest oscilloscopes are able to observe electric vibrations at timescales of just dozens of picoseconds. Many other technologies function at these speeds, such as picosecond lasers, optic fibers and chip technology.

2.1 × 10⁻¹¹ seconds
Quantum vibration in the ammonia molecule

The ammonia molecule, NH₃, has the form of a pyramid. Three hydrogen atoms form the base and a nitrogen atom is at the top (see the drawing of the molecule on the right). The distance between the hydrogen and nitrogen atom in this molecule is 1.014 Ångströms (0.1014 nanometers) and the distance between the nitrogen atom and the base of the pyramid is 0.38 Ångström. In its ground state — with the lowest energy — this atomic configuration according to classical mechanics is stable and the nitrogen atom remains positioned at the same side of the base plane. However, the laws

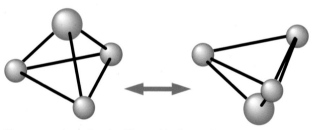

The ammonia molecule. The red balls are the hydrogen atoms, the blue one is nitrogen. Quantum mechanically, the nitrogen atom can tunnel through the base plane to change positions from one side to the other.

of quantum mechanics allow vibrations where the nitrogen atom is sometimes located on one side of the base plane, and sometimes on the other side. To shift positions the atom must 'tunnel' through an energy barrier. Calculations show that such tunneling takes place every 21 picoseconds.

3.3–33 picoseconds
The wave period of radio waves with
extremely high frequencies

Extremely High Frequency radio waves (EHF), with a wavelength of 1 to 10 millimeters, have a frequency of 30 to 300 gigahertz (abbreviated as 1 GHz = 10^9 Hz = 1 billion hertz), and therefore a period of between 3.3 and 33 picoseconds. These waves are used for fixed lines, satellite communications and military objectives, as well as for navigation radars. Above 50 gigahertz these waves are absorbed by oxygen, so they have fewer applications in the atmosphere.

Radiation at this wavelength is generated from all directions in the universe. It originated when the temperature decreased to just below 3,000 kelvins, about 380,000 years after the Big Bang. This radiation has now cooled down to a temperature of about 2.7 kelvins.

Just like radio waves, heat radiation is electromagnetic. Objects with a temperature of one kelvin, that is one degree above absolute zero (−272.3 degrees Celsius), still emit extremely weak radiation, which peaks at about 100 GHz.

Radar on a warship.

**2.3 × 10⁻¹¹ seconds after the Big Bang
A phase transition in the universe**

Up to this moment in time, the electromagnetic and weak forces between particles in the universe were identical, but then a phase transition occurred. Neutrinos and electrons started to differ from each other, and the various types of quarks in the Big Mash (the quark-gluon plasma, referred to at the end of the section about inflation at 10^{-36} seconds and at Chapter 43), started to differentiate.

The top quarks were the first to separate and decay. Important to note is that up to this point, transitions were possible between baryons and antibaryons, so that matter and antimatter could transform freely into each other. But this state of affairs soon got 'fixed'. Apparently, a small surplus of matter particles remained because they always outnumbered the antimatter particles. Originally this difference was extremely small and appeared insignificant — circa one per billion — but every antiparticle sought another particle to convert into pure energy through a process called annihilation of particles and antiparticles, until only a small number of matter particles remained. As such, antimatter disappeared rapidly after annihilation, leaving a universe filled with only matter particles.

Certain forces are known to have been responsible for the small asymmetry that was created between particles and antiparticles, but we are still in the dark as to the details of this process, also referred to as baryogenesis.

At the moment of the phase transition the temperature of the universe was 10^{15} kelvins. The horizon and the visible universe spanned a distance of about 350 million kilometers, and the density of the universe was more than 10^{25} kilograms per cubic centimeter.

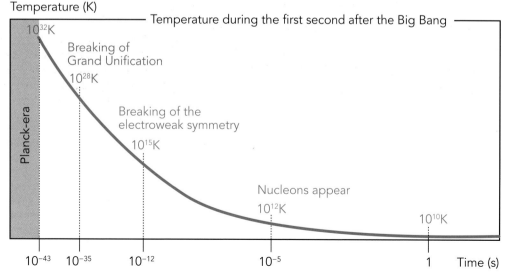

Temperature (K)

——— Temperature during the first second after the Big Bang ———

10^{32}K

Breaking of
Grand Unification
10^{28}K

Planck-era

Breaking of the
electroweak symmetry
10^{15}K

Nucleons appear
10^{12}K

10^{10}K

10^{-43} 10^{-35} 10^{-12} 10^{-5} 1   Time (s)

The evolution of the temperature of the universe. Somewhere between 1 and 10 picoseconds after the Big Bang, the symmetry between electromagnetic and weak forces was broken. The cause of the symmetry breakdown was the Higgs particle, which became active at this time and caused the weak force to function only at short distances (the weak force works only at distances the size of an atomic nucleus), while the electromagnetic force continued to work at extremely large distances. To enable us to positively identify how the electroweak phase transition took place, it was important to identify evidence concerning the existence of the Higgs particle. In June 2012, the two large collaborations of experimental researchers at the Large Hadron Collider at CERN in Geneva announced that a new particle had been detected that was likely to be identified as the Higgs particle.

Chapter 38

10^{-10}

10^{-10} seconds = 100 picoseconds

100 picoseconds is a tenth of a nanosecond. The space between the two arrows below marks the distance light or a radio signal travels within this time.

100 picoseconds is also roughly the time between two collisions of gas molecules. A gas consists of numerous molecules with a particular energy, depending on temperature, pressure and density of the gas. Gas molecules move relative to each other and collide constantly. It is obvious that with higher pressures and temperatures more collisions occur. Between the two collisions molecules move freely, and scientists call the average distance that molecules travel in between collisions their *mean free path* (see figure on the right). For many gases — at 0 degrees Celsius and a pressure of one atmosphere — their mean free path lies between 20 and 200 nanometers. For oxygen this is 63.3 nanometers. If the average speed of molecules is known (which depends on temperature and the molecules' mass), it is possible to calculate the time between two collisions. For oxygen at 0 degrees Celsius and one atmosphere the average speed is 425 meters per second. The *collision time*, the average time between two collisions, is then derived from the ratio between the mean free path and the average speed. In this manner, it can be calculated that the average collision time for oxygen is 149 picoseconds.

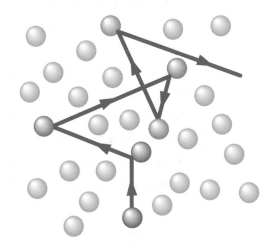

The motion of molecules in a gas. The mean free path is the average distance between two collisions. The time that it takes for two collisions to occur is the average collision time.

1.087,827,757 × 10⁻¹⁰ seconds
The vibration period of the ¹³³Cs-atomic
clock

In Chapter 1 we discussed the second, the fundamental unit of time. This is no longer defined as
1/3,600th part of the 1/24th of a day, because there
are too many variations in the rotational motion
of the Earth. At present, a much more precise
definition has been made possible with the aid of
atomic clocks. As of 1967, the second is defined
as 9,192,631,770 (rounded up to 9.2 billion) vibrations of radiation caused by electrons that decay
between two so-called hyperfine energy levels in
the ground state of the isotope cesium-133.

The principle of an atomic clock is the same for
each atom. Transitions of electrons between two
energy levels cause electromagnetic radiation with
an extremely precise frequency (see image below).

The oldest working
astronomical clock, in
Prague. It is synchronized
with an atomic clock,
which results in extremely
accurate timekeeping.

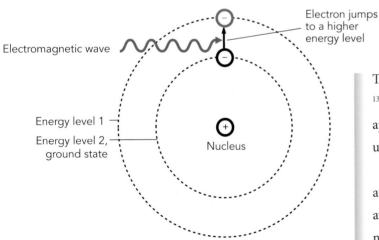

The principle of an atomic clock. Modern atomic
clocks are extremely stable and are off by only one
second every 10 million years.

The period of this radiation is extremely stable for
¹³³Cs, and the duration of 9,192,631,770 vibrations
appears to coincide exactly with the second that
used to be defined as 1/86,400th of a day.

The hyperfine splitting of energy levels in
atoms is caused by an interaction between the
atomic nucleus and the electron cloud. The atomic
nucleus spins around its axis, and the electrons
spin around their axis as well as around the nucleus. All these rotational movements/rotations

cause magnetic fields, influencing each other. The hyperfine vibrations are fluctuations in the relative spin direction of the atomic nucleus and the electron cloud. In essence, this is a precession movement, comparable to the fluctuation of the Earth's rotation axis under the influence of gravity interactions with the moon (see Chapter 14). However, the period of that precession is 26,000 years!

0.33–3.3×10^{-10} seconds
Period of radio waves with super high frequencies

Super High Frequency waves (SHF) are between 3 and 30 gigahertz and have a wavelength of 1 to 10 centimeters. Therefore, they vibrate once every 33 to 333 picoseconds. These waves are used for fixed connections but also for radar. Raindrops are big enough to affect this type of radiation, which is why this frequency plays an important role in weather radar systems.

1.03×10^{-10} seconds
The half-life of the Σ^- baryon

1.82×10^{-10} seconds
The half-life of the Λ baryon

These are baryons with *strangeness*, which means that these subatomic particles contain three quarks, of which one is the strange quark s (the letter s stands for strangeness; according to some, the s was originally meant to indicate 'sideways' and not 'strangeness'). As such, the Λ baryon contains an up quark, down quark and a strange

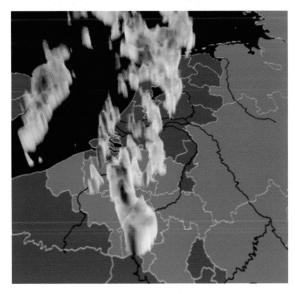

3D image of a rainfall radar.

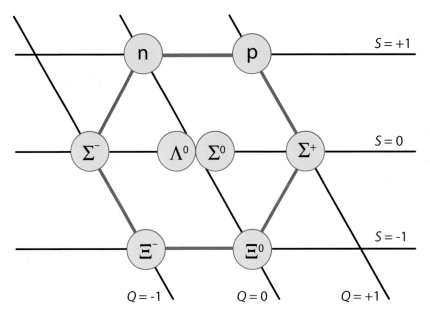

The baryon octet of particles with spin $J = \frac{1}{2}$, arranged according to electrical charge Q and strangeness number S (see also Chapter 26 for similar schematics for mesons). There are also baryons with spin $J = 3/2$, such as the Delta resonance; they occur in what we call a 'decouplet' of ten, of which the charge Q varies between -1 and $+2$, and the strangeness of $S = 0$ to $S = -3$.

quark, which we write as Λ = (uds). Accordingly, for the Σ⁻ baryon this is denoted as Σ⁻ = (dds). The figure on page 149 shows how the various baryons can be arranged according to their charge and strangeness.

During decay the *s* quark transforms into a *u* quark, which creates an intermediary vector boson of the weak force, which in turn decays into leptons or pions. With its mass of 1,115.68 MeV/c², the Λ is heavy enough to decay as follows:

$$\Lambda \rightarrow P + \pi^- \ (64\%).$$
$$\Lambda \rightarrow N + \pi^0 \ (36\%).$$

That the ratio is roughly 2:1 is due to the internal symmetry of the created particles. Because it is electrically neutral, the Λ is not directly visible, but we recognize the traces that the proton and the charged pion leave behind — in the form of the Greek letter lambda (Λ).

The Σ⁻ baryon, with a mass of 1,179.45 MeV/c² decays almost always into a neutron N and a π^-. That this particle decays much faster than the Λ is because of its considerably larger mass surplus.

The K^0_S particle, counterpart of K^0_L (see Chapter 41, 10^{-7} seconds), has a half-life 0.621×10^{-10} seconds. It decays into $\pi^+ + \pi^-$ (69%) or into $2\pi^0$ (31%). That this ratio is roughly 2:1 suggests a similar internal symmetric structure as baryons, although it was much more difficult for theorists to explain this symmetry.

200 picoseconds

About the time a five-GHz microprocessor in a computer requires to add up two integer numbers

The fastest modern computers for sale in 2013 have processors with frequencies of about 5.5 gigahertz, meaning they carry out 5.5 billion calculations per second. These processors contain quartz crystals that function as a clock for the computer — just like the quartz clocks discussed in Chapter 44, but with much higher frequencies. In a single vibration period of the crystal, the processor can carry out a (simple) operation. In principle it is possible to enhance the processor's speed even further, but this would generate a lot of heat and require an extra cooling system.

The first microprocessor, the Intel 4004 from 1971. The whole processor is built on a single chip and contained 2,300 transistors. The clock speed was then only 740 kilohertz.

Chapter 39

10^{-9}

10^{-9} seconds = 1 nanosecond

Nanos is the Greek word for 'dwarf' — *nanus* in Latin. A nanosecond lasts 0.000,000,001 seconds. Light travels 30 centimeters in this time. If you hold this book about 30 centimeters away, it takes a nanosecond for our words to reach your eyes. The nanosecond is also the timescale in which reading and writing processes occur in the memory of a computer with a RAM, or *random access memory*. That is much faster than reading or writing on a hard disk, with timespans of milliseconds, where you must wait for the reader-head to get to the right position first.

0.3 to 3 nanoseconds
Wave period of ultrahigh frequency (UHF) radios

From 300 megahertz to 3 gigahertz, the wave period of a radio signal is 0.3 to 3 nanoseconds. The wavelength is then 10 to 100 centimeters. This wavelength is used for television channels, digital television, satellite communication and the bluetooth in computers and cellphone telephony. Because of the short wavelength, obstacles can easily block these signals.

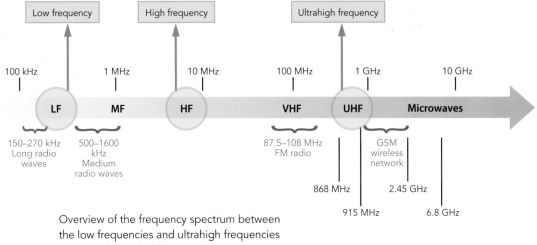

Overview of the frequency spectrum between the low frequencies and ultrahigh frequencies (see also the end of Chapter 35).

The GSM, another word often used in Europe for a cellphone, works at frequencies of between 900 MHz and 1.8 GHz.

GSM, or the *Global System for Mobile Communications*, is in use for digital cellphone telephony in the whole of Europe. It works at frequencies of about 900 MHz and also around 1800 MHz. Outside of Europe other systems are used, around 800 and 1900 MHz. Also the microwave oven works in this frequency range, at 2.45 GHz.

One nanosecond
The period of the so-called 'Lamb shift'

The Lamb shift is a small relative shift between two energy levels of the hydrogen atom. This phenomenon played an important part in the development of atomic physics after the Second World War. In a new formula for electrons, Paul Dirac had taken into account Einstein's relativity theory. With this new formula, the energy levels of the hydrogen atom could be calculated much more accurately.

Nonetheless, small discrepancies remained between theory and experiments, because certain levels — in particular the ones referred to as $^2S_{1/2}$ and $^2P_{1/2}$ — should have shown the exact same energy value. However, a small relative difference between the two was identified. Willis Lamb and Robert Retherford succeeded in 1947 to measure the relative positions of these levels very precisely with the aid of microwaves. The difference corresponds to a frequency of 1,058 megahertz, or one vibration per nanosecond, a surprisingly small effect if you consider the fact that energy levels themselves reach values of 10 electron volts. Via the Einstein-Planck relationship $E = hf$, the frequency corresponds to more than 10^{16} hertz (see the box about Planck's constant, in Chapter 22).

The question as to why these two levels do not coincide was eventually answered by the quantum field theory. The electron emits photons that it then captures again. This causes the electron to fluctuate, and undergo a field strength that deviates a little from Coulomb's law. The effect can be measured very accurately nowadays and is used to test quantum field theory experimentally. Lamb won the Nobel Prize in Physics in 1955 for the *Lamb shift* and the fine structure of the hydrogen atom.

Schematic of the energy levels of the hydrogen atom relating to the Lamb shift. On the left, according to non-relativistic quantum mechanics, in the middle according to relativistic quantum mechanics, and finally, on the right, according to quantum field theory, called quantum electrodynamics (QED). The latter proved to agree with the results of experiments. The shift between the two levels $^2S_{1/2}$ and $^2P_{1/2}$ is called the Lamb shift.

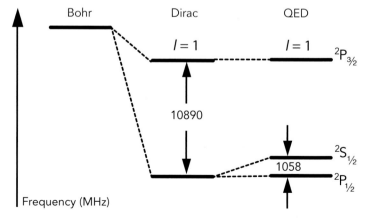

Chapter 40

10^{-8}

10^{-8} seconds = 10 nanoseconds

10^{-8} seconds = 3 meters of light

In a vacuum, light travels a distance of three meters in 10 nanoseconds. Light from a lamp in your living room takes a little more than 10 nanoseconds to fill the room and reach your eye. The speed of light depends on the medium in which it travels. In a vacuum, this is exactly 299,792.458 kilometers per second. In air, the speed is a bit lower, about 299,705 kilometers per second at room temperature at a pressure of one atmosphere. The relationship between the speed of light in a vacuum and another medium is called the index of refraction. For air the refractive index is 1.0003, for glass about 1.5. The speed of light in glass is thus a little less than 200,000 kilometers per second.

$0.3–3.3 \times 10^{-8}$ seconds
Wavelength of radio waves with a Very High Frequency (VHF)

At 30 to 300 MHz the wavelength of radio waves is 1 to 10 meters. These waves have a wide variety of applications, from wireless telephony to FM radio, and some television stations. At the higher frequencies these waves are able to penetrate the ionosphere so that they can be used for satellite systems.

FM stands for *frequency modulation*. This means that a sound signal is first modulated with small variations of the frequency of the carrier wave, instead of the strength or amplitude of the carrier wave, as happens with waves of lower frequencies. A receiver enhances the signal into a set amplitude and then reads the frequency, from which the original sound signal can be deduced, as described in the figure below. The advantage of this is that the strength and stability of the carrier wave signal does not affect the quality of sound as soon as the signal has reached a minimal strength.

Natural light refraction.

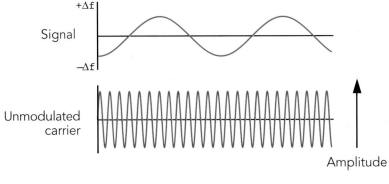

The principle of the frequency modulation (FM) usually provides a better radio reception than the classical amplitude modulation (AM), where the sound signal is modulated with small variations of the amplitude as opposed to frequency.

153

0.858 × 10^{-8} seconds

The half-life of electrically-charged kaons

1.8 × 10^{-8} seconds

The half-life of electrically-charged pions

The charged pion, π^-, has a mass of 139.57 MeV/c^2, and decays into a muon and an antineutrino:

$$\pi^- \to \mu^- + \bar{\nu}_\mu \, .$$

Because the muon behaves in the same way as an electron, you would be forgiven for assuming that the pion could also decay into an electron and an electron antineutrino. The reason why this never or hardly ever happens is because of a characteristic of the particle called helicity. All leptons (such as the electron, the muon and the neutrinos) rotate around their axis: they have spin. When a particle moves at the speed of light, the rotational axis is parallel to the direction of its motion. Helicity then corresponds to the spin direction relative to this axis. It has become apparent that the weak force is only able to produce particles with a left-handed helicity and antiparticles with a right-handed helicity. Because the electron is much lighter than the pion, it should move at almost the speed of light, but then the electron and antineutrino would have an angular momentum of one in natural units while the pion has no spin. This is why the pion decay into an electron is strongly suppressed; because of conservation of angular momentum. This suppression does not occur for the much heavier muon.

For the same reason, the K$^-$ (with a mass of 493.68 MeV/c^2) decays into a muon and an antineutrino, and only extremely rarely into an electron and electron antineutrino. The K$^-$ can also decay into two pions.

Positively-charged particles decay in the same manner: their behavior can simply be deduced by replacing all particles by their respective antiparticles, and vice versa.

25 nanoseconds = 2.5 × 10^{-8} seconds

Time between two collisions of proton bunches in the Large Hadron Collider (LHC)

At 10^{-28} seconds (in Chapter 22) we discussed the Large Hadron Collider in Geneva, a ring-shaped accelerator 100 meters underground that fires protons in opposite directions at almost the speed of light, which then collide. Various bunches of protons, or packets, are prepared and injected into the two rings (see the drawing on page 155). Every bunch contains about 10^{11} protons and is around 10 centimeters long. A maximum of 3,000 packets can be placed into the rings, at a distance of eight meters per packet. The 27-kilometer-long rings are then almost full, and the time between the collisions of two consecutive bunches is about 25 nanoseconds. Actually, when two bunches meet each other, several protons (sometimes more than 20) may hit other protons in the same bunch, and all these collisions are registered; so, the total number of colliding protons can be more than 10^8 per second.

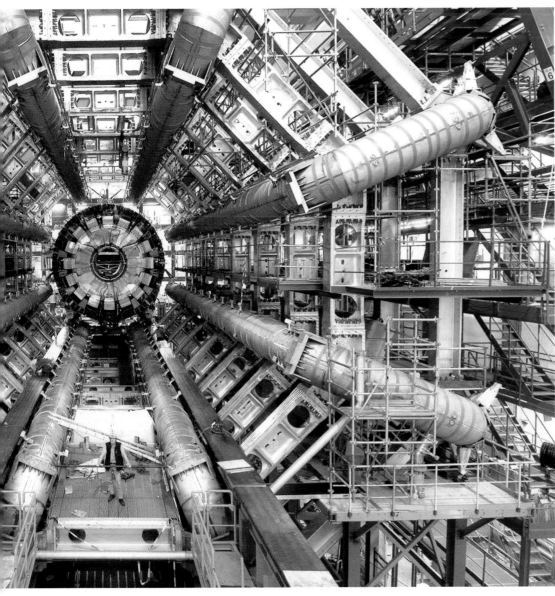

The gigantic dimensions of the ATLAS detector. The protons move in both directions perpendicular to this page (as seen from the readers' perspective, looking at the picture when reading the book), and collide in the middle of the detector.

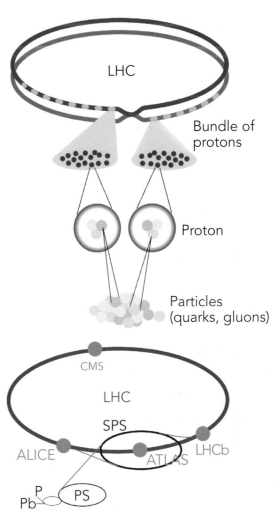

A simplified illustration of the LHC experiment with the proton bunches. Below: a more realistic schematic of the experiment.

Protons are first prepared at lower energies in smaller rings, the PS (26 GeV) and SPS (450 GeV). Then they are injected into the larger ring, and accelerated to an energy of 7 TeV per particle. Four detectors are placed inside the rings: ALICE, CMS, LHCb and ATLAS. At each detector collisions occur and the data is analyzed. The four detectors each carry out their own experiments, partly to study different phenomena and partly to check and verify each other's data.

Chapter 41

10⁻⁷

10⁻⁷ seconds = 100 nanoseconds

10⁻⁷ seconds is 0.1 microseconds. In this time period, light travels 30 meters (in a vacuum). Sound waves travel only 34 micrometers (or 34 × 10⁻⁶ meters) in the same time; that is about a million times shorter. The speed of sound in air depends on the temperature, but we usually assume that we are talking about room temperature. At higher temperatures, the speed of sound is faster.

0.3 to 3.3 × 10⁻⁷ seconds
Wave period of High Frequency
radio waves (HF)

From 3 to 30 megahertz, the wavelength of radio waves is 10 to 100 meters. That is the range of short-wave broadcasting — very popular with radio amateurs. In the lower frequency area, the range increases to around 100 kilometers. Higher frequencies reflect off the ionosphere and enable almost worldwide communication.

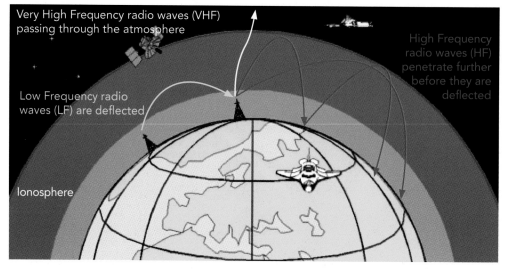

The lower levels of the atmosphere reflect waves with lower frequencies. Higher frequencies penetrate the upper atmosphere before they are also reflected, and therefore have a longer range on Earth. The waves are deflected in the ionosphere, the upper layer of the atmosphere, between 100 and 700 kilometers above the Earth's surface.

0.511×10^{-7} seconds
The half-life of the 'long-living' variant of the K-particles, K^0_L

K-particles or kaons are mesons. Originally, mesons were defined as subatomic particles with a mass between that of an electron and a proton. They can be electrically charged or neutral. Nowadays, all subatomic particles that contain as many quarks as antiquarks (usually one of each) are called mesons.

The neutral kaons have extraordinary characteristics. First, two types of neutral K-particles exist, both with a mass of 497.61 MeV/c². While K^0 consists of a down quark and a strange antiquark, the antiparticle $\bar{K}^0$ contains a down antiquark and a strange quark. However, before they decay, K^0 and $\bar{K}^0$ can mutate into each other. This happens with a frequency of 5.3×10^9 hertz. In turn, this oscillation creates two possible vibration modes, which are named K^0_S and K^0_L. The S and L subscripts stand for *short* and *long*. The two particles are in fact combinations of the original K^0 and $\bar{K}^0$.

Both K^0_S and K^0_L particles behave for 50% of the time as K^0 and the other 50% as $\bar{K}^0$. K^0_S decays into two pions extremely quickly (see Chapter 38, 10^{-10} seconds), but the K^0_L decays much slower, because it must produce three pions to enable decay. This all happens through the weak force. That said, there is an even weaker force that occasionally causes K^0_L to decay into only two pions.

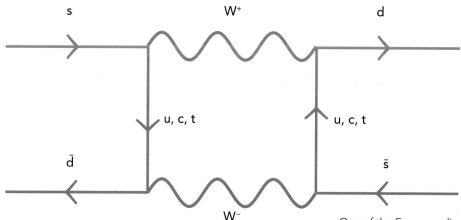

1.09×10^{-7} seconds
The half-life of thorium-218

Thorium-218, with 90 protons and 128 neutrons, has a half-life of 109 nanoseconds. How different the various lifespans of isotopes can be is strikingly evident here: the most stable isotope of this element is ^{232}Th, which has a half-life of 14.05 billion years, a little bit longer than the age of the universe. A remarkable observation can be made about this relatively stable isotope: it could be the energy source of the future. It is technically possible to harvest energy from thorium by bombarding it with protons from a particle accelerator, a procedure that creates relatively few hazardous radioactive waste products — unlike uranium. The world stock of thorium is plentiful. A large part of the heat production in the core of the Earth is attributed to the decay of thorium and uranium. Most thorium isotopes — and ^{218}Th in particular — disintegrate via alpha decay into an isotope of radium, which contains 88 protons.

One of the Feynman diagrams for so-called neutral kaon mixing. On the left, the K^0 meson, which consists of a strange quark and down antiquark. Via the exchange of the charged W boson it mutates into its own antiparticle, the $\bar{K}^0$ meson, which consists of a down quark and strange antiquark. The direction of the arrows for antiparticles is usually indicated as opposite to that of their movement.

Chapter 42

10⁻⁶

10⁻⁶ seconds = 1 microsecond

The timespan of 10⁻⁶ seconds is called a microsecond, or 1 µs, with the Greek letter *mu* as prefix. It is one-millionth of a second, or 0.000,001 seconds. Light travels 100 meters in only 0.33 × 10⁻⁶ seconds (a third of a microsecond). By contrast, the fastest human being in history set the world record for running 100 meters at just below 10 seconds.

One microsecond

The fastest stroboscopes

A stroboscope is an apparatus with which ultra-fast movements are made visible by switching either the observation or the illumination on and off rapidly. The fastest stroboscopes work with lamps that flash a million times per second. This enables us to observe phenomena that last for less than a microsecond. The photograph on the right illustrates the principle of how this works, by showing the movements of a bouncing basketball; but in this case 25 shots per second are sufficient to reveal its movement.

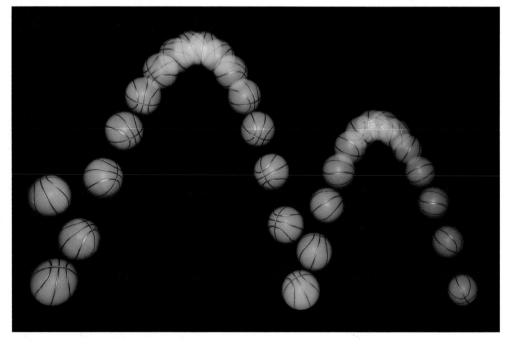

Stroboscopic image of a bouncing basketball.

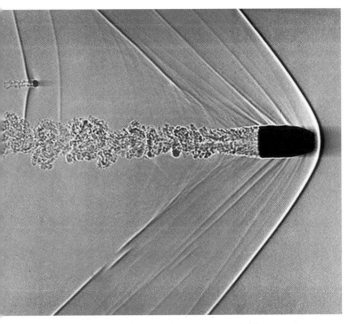

Microsecond image of a supersonic bullet.

1.08 microseconds
The half-life of actinium-218

Various heavy isotopes have a half-life of around one microsecond: ^{218}Ac contains 89 protons and 129 neutrons. It is created when protactinium-222 disintegrates via alpha decay, a process that lasts 3.2 milliseconds. In turn, ^{218}Ac decays as well, resulting in the creation of francium (^{214}Fr). Of the various actinium isotopes, the one with atomic weight 227 is the most stable, with a half-life of 21.77 years. It is a silver-white radioactive material that glows in the dark. Traces of it exist in uranium ore. It can be made by shooting neutrons at radium.

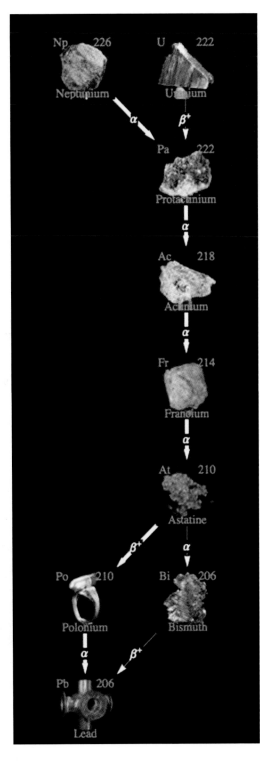

A typical radioactive decay sequence in which actinium-218 appears. Many of these sequences eventually end with one of the stable isotopes of lead, ^{206}Pb, ^{207}Pb or ^{208}Pb.

^{218}Pa 113 µs α			^{221}Pa 5,9 µs α	^{222}Pa 3,3 ms α
^{217}Th 241 µs α	^{218}Th 117 µs α	^{219}Th 1,05 µs α	^{220}Th 9,7 µs α	^{221}Th 1,68 ms α
^{216}Ac 440 µs α	^{217}Ac 69 ns α	^{218}Ac 1,08 µs α	^{219}Ac 11,8 µs α	^{220}Ac 26,4 ms α
^{215}Ra 1,55 ms α	^{216}Ra 182 ns α	^{217}Ra 1,06 µs α	^{218}Ra 25,2 µs α	^{219}Ra 10 ms α
^{214}Fr 5 ms α	^{215}Fr 86 ns α	^{216}Fr 700 ns α	^{217}Fr 19 µs α	^{218}Fr 1 ms α

Various isotopes in the vicinity of actinium-218 and their lifespans. All these isotopes decay via alpha particles (helium nuclei). As the alpha particle contains two protons and two neutrons, the isotope moves down the table two places and also two places to the left.

159

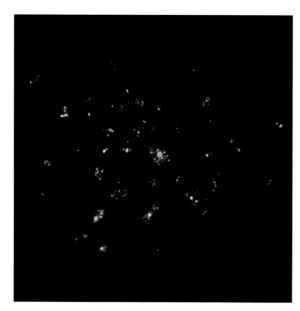

Actinium

**1.523 microseconds
The half-life of the muon**

The muon — denoted by the Greek letter *mu* (μ) — is considered the 'heavier brother' of the electron. Its mass is 206.7 times that of an electron, but otherwise the muon resembles the electron in many respects. Furthermore, all the interaction characteristics of the muon are identical to that of the electron. The muon was discovered in cosmic rays by Carl D. Anderson in 1936. We now know that protons and other extremely energetic particles move towards us from interstellar space. These come into contact with atoms high in the atmosphere, resulting in the production, amongst other things, of pions. The electrically-charged variants of the pions π^+ and π^- are unstable and

decay into a muon and a neutrino (see Chapter 40, 10^{-8} seconds). During its short existence the muon is able to reach the ground, because it does not interact greatly with atoms in the atmosphere. Muons may even penetrate the ground to a depth of a few meters. Eventually the weak force is its downfall: the muon decays into an electron with the same charge and two neutrinos (see figure below).

In subatomic terms, the lifespan of a muon is extremely long, but very short for experimentalists. Future plans exist to make muon colliders, in which muons and antimuons are accelerated, focused into beams and forced to collide. This technical accomplishment would provide unique data regarding subatomic particles, as they provide ideal laboratories to study detailed properties of the Higgs particle. Muon colliders are therefore often called 'Higgs factories'.

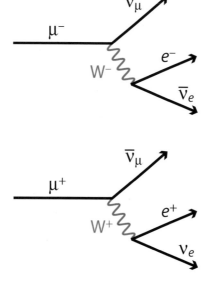

The muon decay. The negatively charged μ^- transforms into a muon type neutrino, indicated by the Greek letter *nu* (ν_μ): the antiparticle μ^+ mutates into an antineutrino $\bar{\nu}_\mu$. This results in the production of a short-lived *intermediary vector-boson*, W, with corresponding electrical charge, which decays into an electron, e^-, and an electron type antineutrino. The antimuon, μ^+, which has a positive charge, decays entirely analogously. The decay products are now a positron, e^+, and an electron type neutrino. Thus, the muon (usually) decays into three particles. A horizontal line above the symbol of a particle indicates its antiparticle.

0.3 to 3 microseconds
The period of radio waves of middle
frequencies

MF (Medium Frequency) periods of 0.3–3 microseconds correspond to 300 kHz to 3 MHz, the frequency of the medium wave. Radio waves of this frequency have a wavelength of about 100 to 1,000 meters. At this frequency they cover a limited radius on the ground. At night, their signals travel much further because of reflection against the ionosphere.

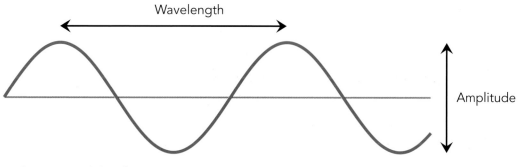

Radio wave with low frequency, thus long wavelength

Radio waves, frequencies and their wavelengths. The figure shows the relation between the frequency and the wavelength. For all waves the formula $\lambda = v/f$ applies where λ is the wavelength, v is wave velocity and f the frequency. For radio and light waves in a vacuum, velocity equals speed of light ($v = c$), and the wavelength is inversely proportional to the frequency. The higher the frequency, the shorter the wavelength — as shown in the figure. As the name suggests, the medium wave lies between the low and high frequencies.

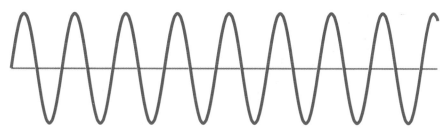

Radio wave with high frequency, thus short wavelength

Chapter 43

10⁻⁵

10⁻⁵ seconds = 10 microseconds

In 10 microseconds, or 1/100,000th of a second, sound travels a distance of 3.4 millimeters. Light, on the other hand, is able to travel almost three kilometers in the same time. The wavelength of radio waves with a frequency of 100 kilohertz is, therefore, three kilometers. The wave period is 10^{-5} seconds.

1–10×10^{-5} seconds
Wave period of ultrasound

Some musical instruments produce sounds ranging up to 100 kilohertz, but the human ear can no longer hear them. Bats can, though, just like whales. Bats use ultrasound as radar. The frequencies of this sound vary enormously, but generally lie between 10 and 100 kilohertz. In 1793, the Italian scholar Lazzaro Spallanzani discovered that bats are able to fly around effortlessly, even when blindfolded. The idea that their hearing was responsible for this feat was only suggested in 1920. Ten years later this was confirmed when microphones were created that made these high frequencies audible to the human ear. Indeed, when the bats' ears were covered up it became apparent that they lost their ability to orientate themselves. Bats produce high frequency beep tones with their snouts, and they deduce what is in their immediate vicinity through the reflection of these sounds. As the wavelengths of these high sounds are a few millimeters to a centimeter, insects and other small objects reflect the waves back to the bat. The position, size and speed of an insect are observed very accurately. Sound as observed by humans is unsuitable for this purpose, since the wavelength of sound waves is too long for that.

10⁻⁵ seconds after the Big Bang
A phase transition in the universe

What did the universe look like only 10 microseconds after the Big Bang? Up to this moment the density was so high – more than 10^{14} kg/cm^3 — that quarks and gluons were still able to move freely amongst each other. We call this 'quark-gluon plasma', or QGP. However, this time period signifies a phase transition, because quarks and gluons had started to regroup themselves in protons and neutrons.

The temperature, at this moment 1.7×10^{12} kelvins, decreased quickly. Even the most distant parts of the universe that we are able to glimpse today with our telescopes were relatively close by at 10 microseconds after the Big Bang: some 250 billion kilometers, or roughly 40 times the present distance from Earth to Pluto.

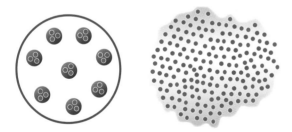

The transition from baryonic matter, such as protons and neutrons, in which quarks are confined, to a quark-gluon plasma. The QGP only occurs at extremely high temperatures or pressure. This state forms something resembling a soup in which quarks and gluons move freely without forming bound states.

Within a fraction of a second, on average 30 microseconds, fate might be sealed when lightning strikes.

3×10^{-5} seconds = 30 microseconds
The average duration of lightning

Lightning is an electrical discharge in the atmosphere, with a rapid succession of lightning flashes and thunder. Usually, lightning strikes somewhere on Earth. This produces an enormous amount of energy in a small area and is combined with extremely high temperatures — in the order of tens of thousands of degrees.

The average output of such a strike is around one terawatt = 10^{12} watt, and that for the duration of about 30 microseconds. Scientists are working on ways to capture and reuse lightning as an energy source — unsuccessfully to date, because machines used to harness lightning incur too much damage as a result of the high temperatures. The exact mechanism of how lightning occurs is complicated and it is difficult to predict where and when it might strike next. Lightning also exists on other planets, such as Venus, Jupiter and Saturn.

$0.33–3.3 \times 10^{-5}$ seconds
Wave period of radio waves with a low frequency

Low frequency electromagnetic waves with a period of about 10 microseconds (30–300 kilohertz, or 1,000–10,000 meters) are indicated as Low Frequency radio waves — LF. This is the kilometer wave, used often for radio broadcasting , especially in Europe. The frequency 40–80 kilohertz (kHz) is also used in many countries to transmit time signals, to which radio clocks are tuned so that they 'automatically' show the correct time. Radio waves at these frequencies can, with favorable weather conditions, follow the curvature of the Earth, so that communication at larger distances is possible. This is partly why radio amateurs are allowed to use an area round the wavelength of 136 kHz.

The LORAN navigational system (abbreviated from *long range navigation*) works at these frequencies. It was developed in the Second World War for sea vessels. It only works with ground stations and has an extremely good range because of the long wavelengths, but nowadays it is scantly used because of the more accurate GPS navigational systems. These radio waves with a low frequency are being applied in long wave broadcasting and inductive systems such as anti-theft portals.

3.9×10^{-5} seconds = 39 microseconds
Time acceleration per day in a GPS clock because of relativity theory

Sir Isaac Newton discovered the laws of nature that determine the movements of the Moon and the planets in our solar system. His theory, published in 1687, is now called classical mechanics. According to this theory, time is an absolute quantity that always passes with the same speed everywhere. With two revolutionary theories, Albert Einstein made an important modification. His first theory, the special relativity theory of 1905, held that time passes differently for different observers. In the eyes of a stationary observer, the time of a fast-moving clock appears to pass slower. Einstein famously derived this from the fact that the speed of light is always the same for every observer, which is different from what would be concluded from Newtonian mechanics. In Einstein's second theory, the theory of general relativity, gravity was taken into account as well, from which Einstein concluded that clocks in a gravitational field run slower than clocks at great heights where

Galileo, a satellite from the European Space Agency (ESA), used for telecommunications and GPS (*Global Positioning System*).

gravity is weaker. A satellite in space moves at high velocity and is located at great altitude (a GPS satellite for instance is 20,200 kilometers high), so that both these effects occur.

GPS satellites have clocks on board, which are important for determining the position of objects on Earth, for example a car on a highway. The position can be calculated by determining the distance between the receiver (the car) and at least three different satellites. The distances are calculated by measuring the time difference between the moments of emitted and received radio signals. As the GPS satellite moves in an orbit around the Earth, the clocks incur a slight delay caused by the movement on the one hand (11 microseconds per day) and acceleration because of decreased gravity on the other (50 microseconds per day). In total there is a time-acceleration of 39 microseconds per day, and it is important that this difference is processed accurately, so that the correct positioning data is sent to the receiver — your car — on Earth.

Chapter 44

10^{-4}

10^{-4} seconds = 100 microseconds = 0.0001 seconds

10^{-4} seconds is a tenth of a millisecond. Light can travel about 30 kilometers in 0.0001 seconds. The protons in the 27-kilometer long tunnel of the Large Hadron Collider at CERN that are accelerated almost to the speed of light take 0.9×10^{-4} seconds to complete one round. In comparison, sound moving through air travels only 34 millimeters in 0.0001 seconds. The wavelength of sound at 10 kilohertz is thus 34 millimeters, and the sound waves with that frequency vibrate once in 10^{-4} seconds. For the elderly among us, sounds with this frequency cannot be picked up anymore. Young children are able to hear pitches of up to 16–20 kilohertz.

0.305×10^{-4} seconds
The period of quartz clocks (32,678 hertz)

Quartz is silicon oxide in crystalline form, a substance that exists in abundance naturally. Pure quartz crystals are easy to produce and stay very stable. The reasons for using them in clocks are twofold: first, the frequency of the crystal depends only on its shape, and not (or barely) on temperature, and second it is possible to use piezoelectricity to convert mechanical stress or pressure into electric vibrations ('piezo' comes from the Greek *piezein*, which means to squeeze or

to press). By adding pressure to certain crystals an electric voltage can be created; the small pressure waves caused by these vibrations are sufficient to generate detectable electric oscillations. Crystals are shaped into tuning forks and are resonated to a frequency of 32,768 vibrations per second (=32,768 hertz). This is exactly 2^{15} vibrations, which is easily converted by an electronic circuit into a signal that drives the hands of a clock. As there is a slight temperature dependency, a quartz clock may run a fraction of a second off. In modern clocks, adjustments are made automatically depending on readings from an internal thermometer. The first quartz clocks, manufactured back in 1927, were bulky machines because of the vacuum tubes that were needed to house the electronics. The first wrist watches based on quartz crystals were sold in 1969.

1.13×10^{-4} seconds = 0.113 milliseconds
The half-life of protactinium-218

With 91 protons in its nucleus, protactinium is positioned just before the element uranium in the periodic table of elements. None of its isotopes is stable. The most stable one — ^{231}Pa — has a half-life of 32,800 years. With 13 fewer neutrons, the half-life of the isotope with atomic weight 218 is considerably shorter at just 0.113 milliseconds. Just like most other lighter protactinium isotopes, it decays by the emission of an alpha particle (helium nucleus). This results in the creation of ^{214}Ac, or actinium-214, which is also extremely unstable.

Above: the first Swiss quartz clock was fabricated just after the Second World War. *Right*: the inner mechanism of a quartz clock. By building in a temperature sensor an even greater accuracy can be obtained.

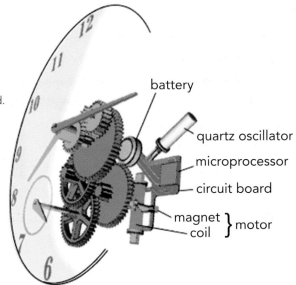

battery

quartz oscillator

microprocessor

circuit board

magnet

coil } motor

3.3×10^{-4} seconds
Period of radio waves with a
very low frequency, VLF

Electromagnetic waves with a period of approximately 0.1 milliseconds (3–30 kilohertz, wavelength of 10–100 kilometers) are indicated with VLF, the abbreviation of Very Low Frequency. This range is better known as the *myriameter* band; the word 'myriameter' is sometimes used to indicate 10,000 meters (or 10 kilometers). For radio broadcasts this frequency is rarely used because there is too little space on the band. Nevertheless, it is useful for a few specific purposes, such as communication with submarines, (provided they are not too deep), or to provide the time. As the waves are transmitted smoothly between the Earth and the ionosphere, the range is global and very stable.

0.0005 seconds = 5×10^{-4} seconds
The rotation period of an ultracentrifuge

With an ultracentrifuge, an object can be exposed to extremely high accelerations. At present, speeds of 120,000 revolutions per minute or 2,000 per second are reached routinely. The principle is similar to the centrifuge in a washing machine (the appliances even look alike from the outside). The acceleration that is created within the walls of the ultracentrifuge reaches up to 625,000 times the acceleration of the Earth's gravity — in short, 625,000 g.

There are centrifuges that create an even larger gravitational field, but those contain gas instead of fluid. A well-known example is the Zippe centrifuge that is used to separate isotope mixtures, for example in the enrichment of uranium. The isotopes ^{235}U and ^{238}U have only a small weight difference, and an artificial gravity field of a staggering one million g is required to separate them. In the Netherlands, this technique is regularly used to produce enriched uranium.

Urenco, gas-ultracentrifuges for uranium isotopes.

Chapter 45

10^{-3}

10^{-3} seconds = 1 millisecond = 0.001 seconds

10^{-3} seconds, also called one millisecond, is the time that light and radio waves need to cross a distance of about 300 kilometers. Sound travels only 34 centimeters in this time. It is also the wavelength of the high B (the high 'si' in music, see the figure on the right) in air at room temperature. The fact that sound is like a wave means it can easily bend itself around obstacles. But a wall with much larger dimensions than the wavelength may well be soundproof, especially for higher tones, so barriers are often used to absorb traffic noise alongside highways.

The musical note B5

About one millisecond is also the duration of the flash on an electronic camera. Modern cameras have a built-in flash. In a tube filled with xenon gas, a high electric charge is used to generate an arc discharge, which creates a bright light flash of about 1/1,000th of a second. It is also possible to create much shorter light flashes — to about 50 μs = 50×10^{-6} seconds. Obviously the camera's shutter must be lined up precisely with the flash. The shutter is mechanical and usually consists of two curtains: a top one that moves across the focal plane, with a second that follows behind, effectively moving a slit across the focal plane until each part of the film or sensor is exposed. The width of the slit that is created by the two moving curtains is the measure of the shutting time. As the whole lens must be opened for the flash, the shutter speed is usually limited to about 1/200th of a second. If this time needs to be shortened even further, other shutter techniques can be used.

> $0.00089 = 0.89 \times 10^{-3}$ seconds
> The half-life of ununoctium-294, a super heavy element

Around 1950, nuclear physicists raised the supposition that the protons and neutrons comprising atomic nuclei are able to form relatively stable configurations at extremely high values of the atomic number (the number of protons in its nucleus). While it was known that elements with an atomic number above 100 are unstable, it was believed that those with the numbers 114 and 126 might have a lifespan of billions of years.

Around 1994, German researchers succeeded in creating a few very heavy elements, by colliding the nuclei of lighter elements, namely 110, 111 and 112. To date, atomic number 118, ununoctium, is the heaviest element that was identified. So far, only three or possibly four atoms were produced and detected, by bombarding californium-249 with ions of calcium-48. The most stable form of element number 118 has a half-life of 0.89 milliseconds. Ununoctium is grouped under the noble gases. It decays into element 116 with the emission of an alpha particle.

A longer-living super-heavy element is the one with atomic number 114 ('ununquadium', but more recently given the name 'flerovium'). It has been produced in larger quantities and its most stable isotope has a half-life of 2.6 seconds. The names of these super heavy elements are temporary; the lineup of the three digits in the atomic number were translated in mock Latin. Sometimes the name of the lighter element that is positioned directly above the new element in the periodic system is used, with the prefix 'eka'. Below is a short table with the half-lives of a number of super heavy elements.

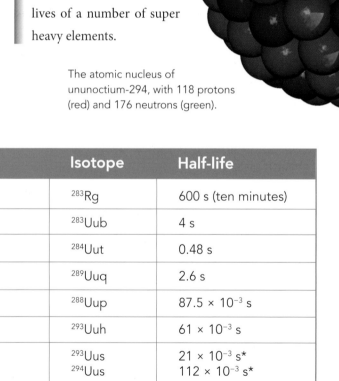

The atomic nucleus of ununoctium-294, with 118 protons (red) and 176 neutrons (green).

Atomic number and name		Isotope	Half-life
111	Roentgenium = unununium or eka-gold	^{283}Rg	600 s (ten minutes)
112	ununbium or eka-mercury	^{283}Uub	4 s
113	ununtrium or Uut	^{284}Uut	0.48 s
114	ununquadium or Uuq	^{289}Uuq	2.6 s
115	ununpentium or Uup	^{288}Uup	87.5×10^{-3} s
116	ununhexium or Uuh	^{293}Uuh	61×10^{-3} s
117	ununseptium or Uus	^{293}Uus ^{294}Uus	21×10^{-3} s* 112×10^{-3} s*
118	ununoctium or Uuo	^{293}Uuo ^{294}Uuo	0.12×10^{-3} s 0.89×10^{-3} s

*These values were determined only recently, and are far from certain.

The musical note B5 (a 'si' at the fifth octave) has a frequency of 987.8 hertz, or a vibrational period of 0.00101 seconds. Only a soprano can reach this note — and maybe even one note higher, up to a C6. A mosquito moves its wings back and forth 275 to 575 times per second, in some cases even 1,000 times per second; its sound would reach a very high note, though the maximum is a high B (see the music stave on page 169).

Electromagnetic waves with a period of about one millisecond (one kilohertz, wavelength of 300 kilometers) are called ULF waves (Ultra Low Frequency). Such waves exist naturally in the Earth's magnetosphere. These have applications in mining, because the waves are able to penetrate the ground very deeply. They are also used in communication between submarines, as they travel easily through water over long distances.

0.0014 seconds
Millisecond pulsars

Pulsar J1748-2446ad, discovered in 2004, is the fastest-spinning neutron star known to man (2011). Like most millisecond pulsars, this one is located within a spherical cluster of thousands of stars. The distance to the Earth is about 18,000 light-years. It rotates around its axis an impressive 716 times per second. It is possible to determine its rotational period very precisely: one rotation in 0.001,395,954,82 seconds. This star is almost twice as heavy as the Sun, but has a diameter of only 32 kilometers. Along its equator the rotation speed is almost a quarter of the speed of light. The high rotation speed must be caused by the absorption of matter from a companion star. Its companion has a mass of 0.14 solar masses but is larger than the Sun and spins with an orbital period of about 26 hours around the pulsar, which causes eclipses.

Another extremely fast spinning pulsar is J1903+0327. This one achieves 465 rotations per second. We will devote more attention to this pulsar in the last chapter of this book.

In 1999, a small star was found that may be spinning more than a thousand times per second. ESA's gamma-ray observatory found that the neutron star XTE J1739-285 may be spinning at 1122 Hz, but further study here is still necessary to confirm this observation. This star must be extremely compact.

0.002 seconds = 2 milliseconds
Functioning of nerve cells

Nerve cells, or neurons, are the information processors in our body. Through chemical reactions they are able to receive and emit signals. The neurons in our nervous system — we have an estimated 100 billion — form a complicated electric circuit that conducts information to all parts of our body. Nerve cells are interlinked via a thread-like structure, the axon (or neurite), which can be up to one meter in length. They form neural networks with up to thousands of neurons.

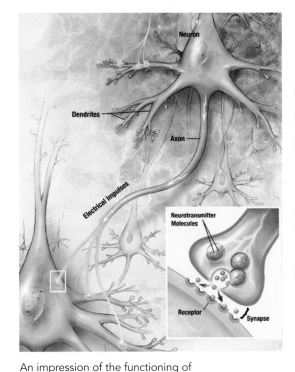

An impression of the functioning of our nerve cells or neurons. In the branches of neurons, the so-called dendrites, signals arrive. This excites the cell bodies (the widened ends of the nerve cell), firing off signals to neighboring cells via the axons. The ends of the axons branch out, and via so-called synaptic connections the branches make contact with the dendrites of neighboring cells. In a synapse, the transmission of impulses takes place biochemically through neurotransmitters.

By a change in the electric voltage on the cell membrane, a neuron can be activated or excited: it is able to receive a signal and forward it to another cell via the axon. This process is comparable in some ways to an electric discharge, which is why it is often said that a neuron 'fires off' information.

The whole process takes only two milliseconds. It is followed by a period of rest, after which the cell is reset to its starting position and can be excited again. Not all nerve cells are equally active; some fire dozens of times per second, others are in a quiescent mode for minutes on end. Immediately after a signal is fired off, the propagation velocity is between 1 and 100 meters per second, depending on the characteristics of the nerve fibers. This propagation velocity determines the speed of our physical reactions (see Chapter 47).

A question of milliseconds
The fastest gamma ray bursts in the universe

Every day, special observation satellites detect strange bright flashes of gamma rays (see Chapter 28) in space, coming from all directions. Elaborate research has determined that these gamma ray bursts originate from far away galaxies. The origin must be exploding stars, but the explosion processes appear to vary each time. The pulses often last from a few seconds to a few minutes, but the fastest flashes deposit their energy in mere milliseconds. The slower pulses probably originate from extremely violent supernova explosions, where the star's nucleus transforms into a black hole. The fastest gamma ray bursts are likely to come from the fusion of two compact stars or black

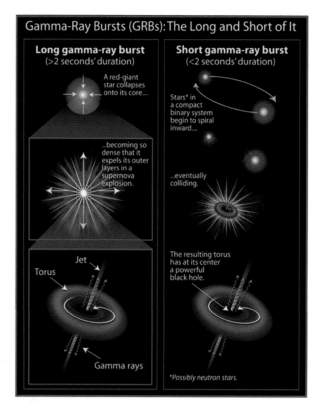

The two different mechanisms of gamma ray bursts, the short and long gamma ray bursts (GRB).

holes into one black hole. The emitted radiation is believed to be concentrated into narrow beams. We only detect those beams that happen to be aimed in our direction, so the total energy emitted by the source that we observe might actually be less than it looks; looking into a torch directed at us makes the light appear very bright, brighter than the light of an ordinary light bulb, even though the latter may emit more light in total. But even so, the amount of energy released in these short bursts is comparable to the total amount of light radiated from our Sun in its entire existence of ten billion years.

Chapter 46

10^{-2}

10^{-2} seconds = 10 milliseconds = 0.01 seconds

We have arrived at 1/100th of a second, or 10 milliseconds. These are timescales that are important in speech technologies and speech recognition. It is the difference in time between the pronunciation of 'sometime' and 'some time'. Another example is the shutter speed of cameras, which last around 1/100th of a second, depending on the amount of available light, the chosen aperture opening (the f-stop value) and the sensitivity of the film or sensor. As an aside, light travels almost 3,000 kilometers in 0.01 seconds.

0.009965 seconds = 0.9965×10^{-2} seconds
The half-life of ^{13}N

This nitrogen isotope transmutes into the stable carbon isotope ^{13}C with a half-life of 9.965 milliseconds. The isotope can be created when the most common oxygen isotope, ^{16}O, is bombarded with protons:

$$^{1}H + {}^{16}O \rightarrow {}^{13}N + {}^{4}He$$

In the interiors of heavy and hot stars this nitrogen isotope plays a very important role in nuclear reactions, where hydrogen is converted into helium, the so-called CNO cycle. By capturing protons, a normal ^{12}C atomic nucleus transmutes

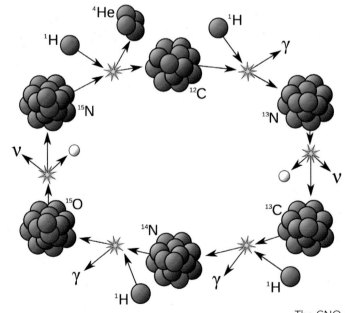

	Proton	γ	Gamma Ray
	Neutron	ν	Neutrino
	Positron		

The CNO cycle occurs with nuclear reactions in heavy stars, where hydrogen is converted into helium. Positrons, neutrinos and gamma radiation are released as well during this process.

into ^{13}N, which beta-decays into ^{13}C. As ^{13}C captures another proton and is transformed into ^{14}N, ^{14}N becomes ^{15}O, which decays into ^{15}N, which mutates into ^{12}C + ^{4}He through proton capture. Thus, the carbon nucleus returns, but as part of this process four protons are transformed into a helium nucleus and two positrons (the latter appeared during the decay processes). This shows that carbon (C) works as a catalyst for the reaction 4H → He. The helium nuclei that are created through this process will get the opportunity to fuse much later on, to form heavier elements.

The CNO cycle is the most important process in which hydrogen is converted into helium in stars that are considerably heavier than the Sun. The helium concentration increases starkly because of it, after which the formation of heavier elements from helium starts. If such a heavy star explodes, heavier elements enter interstellar space and a new generation of stars and planets that contain heavy elements can come into existence, such as the Sun and the Earth.

0.01 second = 1×10^{-2} seconds
100 hertz

The musical note G2 (a 'sol' on the second octave) has a frequency of 98 hertz, or a vibrational period of 0.0102 seconds. A baritone has a range from G2 through two octaves to G4.

Electromagnetic waves with a period of about 0.01 seconds (100 hertz, wavelength 3,000 kilometers) are indicated as SLF (Super Low Frequency). Presumably, the Earth emits powerful radio waves of 50 and 60 hertz into space, originating from our many high voltage cables. Electromagnetic waves with periods between 0.03 and 0.3 seconds (3–30 hertz) are indicated as ELF (Extremely Low Frequency). Their wavelengths therefore vary between 10,000 and 100,000 kilometers.

0.02 seconds
50 hertz

Two-hundredths (2/100th) of a second is the period of the alternating current of 50 hertz with which electricity is distributed in Europe. The reason for using this frequency is because heavy turbines in power stations have this rotation frequency, and alternating currents are technically much easier to transform from higher to lower voltages or vice versa than a direct current. The United States uses a frequency of 60 hertz — a period of 0.0167 seconds. In some countries, such as Japan, both frequencies are used. For certain applications lower frequencies are in use, such as 16⅔ hertz by railways in some countries.

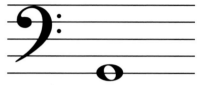

The musical note G2

This hummingbird moves its wings about 50 times per second to hover in the air like a helicopter. The picture was taken with a short shutter speed so that there is hardly any motion blur.

0.0125 – 0.066 seconds

The duration of a wing beat of a hummingbird

The hummingbird is one of the most diminutive bird species — the smallest weighs even less than two grams. With its long, fine beak and whip-thin, split tongue, the hummingbird is equipped for sucking honey from flowers. In turn, some types of flowering plants depend on the hummingbird for pollination. To be able to hover in the air, hummingbirds have developed an exceptionally fast wing beat, with a frequency of 15 to 80 strokes per second.

0.010378 seconds

The rotational period of pulsar B1639+36A

Pulsar B1639+36A is located in star cluster M13, around 24,000 light-years away, and turns once around its axis in 0.010378 seconds. Not a lot is known about this pulsar; it appears to be very old, probably billions of years.

A normal house fly (*Musca domestica*) beats its wings 200 to 300 times per second.

Chapter 47
10⁻¹
10⁻¹ seconds = 100 milliseconds = 0.1 seconds

Our reaction time is at least 0.1 seconds. This can be measured with an array of experiments, for example, having someone push a button in response to a signal. This is called a simple reaction. The response time with a simple reaction depends on various factors: we tend to react faster to sounds (on average within 0.16 seconds) than to visuals (on average 0.19 seconds); and the response time is age dependent as young people tend to have faster reflexes than the elderly.

Another sort of experiment is a 'go/no-go' reaction. This involves responding by pressing a button when a red picture is shown and not pressing a button with a green one. The response time is then somewhat delayed because the signal (color of the picture) must be interpreted first. Average response times then take around 0.3 seconds.

A third type of reaction is the 'choice' reaction, where the button must be pressed with the middle finger at a red picture and the index finger at a green picture. Average response times are then around 0.4 seconds. Generally, a response time consists of two or three parts: the time required to identify the signal, (possibly) the time to make a choice as to how to respond, and then the time it takes to carry out the movement.

One of the pioneers of this research was the physiologist and medical practitioner Franciscus Cornelis Donders (1818–1889). Donders, with a few of his followers, was one of the founders of mental chronometry, the specialization in experimental psychology that focuses on determining cognitive processes and reaction times.

F.C. Donders was also a well-known ophthalmologist. The terms 'far-sighted' and 'near-sighted' were coined and studied by him.

0.05 seconds or longer
The period of infrasound

Infrasound (or infrasonic sound) has a frequency of less than 20 vibrations per second. A period therefore lasts at least 0.05 seconds. That is roughly the lower limit of the human ear. Infrasound can be created in many ways, such as by breaking waves, avalanches, earthquakes or thunderstorms. The Comprehensive Nuclear Test Ban Treaty Organization uses infrasound detectors, among other things, to detect secret nuclear explosions. Airplanes that cross the sound barrier produce infrasound that can be detected at great distances as well.

Various different animal species are known to be able to hear much lower frequencies than humans: whales, elephants, rhinos, and many more use infrasound to communicate with each other. Elephants, for example, can pick up infrasound from their herds hundreds of kilometers away.

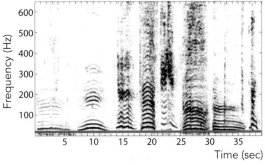

Registration of the blaring of a forest elephant, taped at the Dzanga clearing in the Central African Republic by a research team studying elephant sounds. The red line indicates the lower limit of the human ear. Below that sits infrasonic sound. Elephants also make many sounds that we humans *are* able to hear.

The beluga whale (*Delphinapterus leucas*) is a white whale that communicates with its own species at great distances with the use of infrasound.

A wink takes about 50 to 80 milliseconds. The eye is unable to distinguish sequential images that scroll faster than 30 times per second. A television signal generally contains 25 to 30 images per second. To create a steady picture without too much flickering, a modern television converts this to 50 to 100 images per second. Furthermore, a television signal of 25 to 30 images per second does not transmit the information for a completely new picture, because a signal cannot contain this much information; that would require a bandwidth of 200 to more than 800 megahertz. The reason this is not a problem is that a normal television picture always contains parts that are the same and because elements of images usually change only slowly, such a signal can often be strongly compressed. This happens frequently. A television signal does not usually not contain more than 20 to 50 megabits per second. As an aside, there are many different broadcasting systems in use in the world.

0.1 seconds
The half-life of ^{32}Ar (0.098 seconds) and ^{24}Si (0.102 seconds)

Argon is a noble gas that normally consists of 18 protons and 18, 20 or 22 neutrons. With only 14 neutrons, ^{32}Ar is unstable. By electron capture or positron emission it converts into the chloride isotope ^{32}Cl, which changes into sulfur ^{32}S, which is stable.

Just over a quarter (25.7%) of the weight of the Earth's crust consists of silicon. It reacts with oxygen to form silicon oxide, SiO_2, a basic component of sand, quartz and many other rocks. Silicon has atomic number 14, which means that its nucleus contains 14 protons. The number of neutrons in stable silicon is 14, 15 or 16.

But isotope ^{24}Si contains only 10 neutrons and is very unstable. In 0.102 seconds, the nucleus captures an electron from its surroundings or emits a positron, which in both cases is accompanied by the emission of a neutrino. This causes the nucleus to change into the aluminum isotope ^{24}Al with 13 protons and 11 neutrons, which in turn decays within two seconds into stable magnesium-24. The isotope ^{24}Si does not exist naturally, of course, because of its short half-life, but can be created by bombarding lighter nuclei with protons.

0.104182 seconds
The rotational period of pulsar J1811-1736

Pulsar J1811-1736 was one of the first pulsars to be discovered. Its rotational period could be determined very precisely because of its pulsing signal: it turns around its axis once in 0.104182 seconds. As a result of the small variations in its signal, it was discovered that this pulsar moves around another compact object in an eccentric orbit, once every 18 days and 19 hours. The masses of the two companions are similar and together they amount to more than 2.5 solar masses. This was determined by the detection of the perihelion precession, an effect on the orbital motion that can only be understood as a result of Einstein's relativity theory (similar to the precession of the perihelion of Mercury, which we discussed in Chapter 9). It is likely that the pulsar's companion is also a neutron star.

0.1 seconds

Fast musical notes

Musicians can play dozens of notes per second, even as many as 50 per second in a glide known as a *glissando*. In musical notation, the length of a note is indicated by the number of flags attached to it; every flag indicates that the note is shorter by a factor of two. The common notation for the duration of notes is provided in the image below.

0.12 seconds

The time a radio signal requires to reach a geostationary satellite from Earth

Many communication satellites are located in a geostationary orbit, at a height of exactly 35,786 kilometers above the equator. So the signal of a telephone call must travel twice that distance to reach its goal and therefore takes at least 0.24 seconds. That means that, during a conversation via satellite, we have to wait twice that time — half a second — for an answer.

In this orbit, the satellite turns at the same speed as the Earth, so that the object, when viewed from the ground, always appears in the same position. With just a handful of such satellites, it is easily possible to maintain communication channels that cover the entire Earth.

0.133 seconds
The time a radio signal requires to orbit the Earth

0.32 seconds after the Big Bang

After less than half a second, neutrinos detached themselves from other particles. Neutrinos always interact with other particles via weak interactions, but the weak force is losing its grip on the universe at this moment. Up to this point their temperature was the same, 1.6×10^{10} kelvins, but the neutrinos' temperature lowered a bit faster than that of other particles — the latter received heat from the electron-positron annihilation that takes place now. The universe continues to be filled with a neutrino background, but it is even weaker and more difficult to detect than the background radiation (2.7 kelvins) of photons.

The size of the universe was at this moment just beyond four light-years, which is about the current distance to the nearest star outside our solar system. At 0.32 seconds after the Big Bang, the density was still 100,000 kilograms per cubic centimeter.

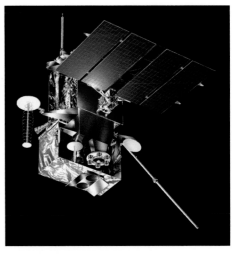

A geostationary satellite, in this case used to determine accurately locations on Earth.

The duration of musical notes.

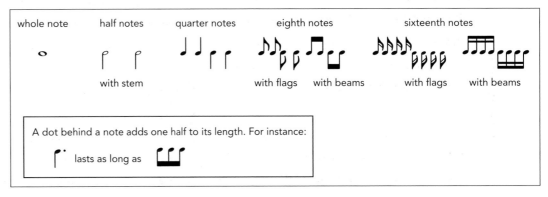

whole note	half notes	quarter notes	eighth notes	sixteenth notes
	with stem		with flags	with beams	with flags	with beams

A dot behind a note adds one half to its length. For instance: ♩˙ lasts as long as ♫♪

Chapter 48

10⁰

10⁰ seconds = 1 second

In the final chapter, we come full circle back to where we started: the second. A second is the time period that we relate to the most — a heartbeat lasts a second when we are in a state of rest. It is also roughly the time you need to say the word 'Amsterdam'. Of course, the second is the official scientific unit of time.

0.77 seconds

One revolution of a 78-rpm record

The turntable was the most frequently-used machine to listen to recorded music until about 1980. Thomas Alva Edison invented the phonograph about a century earlier, in 1877. It worked with cylinders and was the first apparatus that could record and reproduce sound; the much more popular gramophone was based on

record discs and was introduced and patented by Emile Berliner in 1888. They got into large industrial production from 1918 onwards, after the patent term was finished. The oldest record discs span at a rate of 78 turns per minute — that is one turn in 0.77 seconds — but as the quality improved, new speeds were introduced: 45, 33 and 16 rpm. Sixteen rpm meant one revolution every 3.6 seconds.

Thomas Edison's cylinder phonograph

Emile Berliner's gramophone

In the 1980s, the CD-player became popular and displaced the turntable. A CD, or compact disc, turns much faster than a gramophone and makes dozens of rotations per second, depending on the position of the record.

One second in music

A metronome is an instrument that ticks to indicate the desired speed of a musical piece. Differing types of music are played at various tempos. Sometimes a composer clearly denotes the speed at which music should be played, but mostly it is left to the interpretation of the performer. That said, guidelines do exist, which are usually indicated with Italian phrases. Below is a short overview of common tempi:

Largamente: 10 ticks per minute

Larghissimo: 20 ticks per minute or less

Largo or Lento: 40–60 ticks per minute

Adagio: 66–76 ticks per minute

Andante: 76–108 ticks per minute

Allegro moderato: 112–124 ticks per minute

Allegro: 126–168 ticks per minute

Presto: 168–192 ticks per minute, about 3 ticks per second

Prestissimo: 200–208 ticks per minute

Sometimes the length of an eighth of a note is indicated, and sometimes those of quarter notes (see the previous chapter). In modern electronic dance music 'bpm' is used, or beats per minute. The musical genre 'hip-hop' falls into the category 85–120 bpm, and 'house' plays at 110–140 bpm.

One light-second

Light travels exactly 299,792,458 meters in one second, about two-thirds the distance to the moon (see Chapter 1). A light-second is thus almost 300,000 kilometers. In 1983, the Conférence Générale des Poids et Mesures (CGPM) determined that this defines the meter as a unit of length. Previously, the length of a bar of platinum was used as a 'standard meter', or it was defined by the wavelength of an atomic spectral line.

Sound travels almost a million times more slowly than light. In air at room temperature, sound travels about 340 meters in one second. By counting the number of seconds between lightning and thunder, it is fairly easy to determine the distance to the storm: about 340 meters for each second.

Fast-rotating stars

Pulsating stars — pulsars in short, also referred to as neutron stars — are extremely compact and fast-rotating stars with a strong magnetic field (see the figure on the next page). They started their lives even larger and heavier than the Sun, but in their current state their nuclear fuel has almost completely burned out. This caused them to shrink significantly and rotate quickly. Their rotational period may have decreased further when the matter from a nearby star was sucked up. The pulsars' mass was still larger than that of the Sun, while their size was reduced to only a few kilometers. This means their density must have increased to hundreds of millions of tons per cubic centimeter, the density of matter in a nuclear atom. Their weight is even millions of times higher, because gravity at the surface is so much stronger than on Earth. In fact, the counter pressure of the matter is only just strong enough to prevent further collapse into a black hole.

The 'nuclear matter' of which pulsars are composed consists almost entirely of neutrons; this is because the only other stable nuclear particle, the proton, is electrically charged. Because of the repulsive electric forces between the protons, their charge should be compensated by electrons, but these are much less easily condensed into a small volume with such a high density. This is a quantum mechanical feature of electrons; due to their low mass, they are associated with quantum waves that vibrate slowly and therefore have long wavelengths. Electrons must stay out of each other's way by at least one wavelength. So the protons are forced to transmute into neutrons or else they are pushed out of the system. Thus, the most compact stellar objects are almost entirely made out of neutrons.

The axis around which the pulsar pivots has a certain angle relative to the direction of its magnetic field (just as with the Earth, where the geometric pole and magnetic pole are not one and the same). As a lot of radio radiation is emitted precisely at the magnetic poles, it appears to pulsate — just like the beam from a lighthouse. The frequency of the radio pulses corresponds with its rotational velocity. This turns out to be very stable and can vary between one rotation per millisecond (0.001 seconds) to almost ten seconds. So pulsars with a rotational period of one second spin relatively slowly around their axis.

The pulsar has a strong magnetic field of about 10^8 tesla. To compare: our Earth has a magnetic field that varies between 1 and 6×10^{-5} tesla — so about thirteen orders of magnitude less than a pulsar. MRI scanners in hospitals use magnets with a strength of about 2 to 7 tesla. The strength of magnetic fields that are used in particle accelerators, such as at CERN in Geneva, is about 8.33 tesla. The strongest magnetic field ever made by man was about 100 tesla. These strong fields can only be maintained for about a second.

The theory of pulsars is not yet well developed or understood. When they were first discovered, they were given the name LGM — for *little green men* — because it was as if the Earth was receiving signals from extra-terrestrial life from outer space.

It has been estimated that more than 100,000 pulsars exist in our Milky Way. Below are two examples with a rotational period of about one second.

1.004,037 seconds
The rotational period of pulsar B1718-19

This pulsar is part of a binary system, but instead of double stars circling each other, one partner is a pulsating neutron star. The corresponding star would have a mass of about 0.1 to 0.2 solar masses. This pulsar was discovered in 1993 and is believed to lie in the global cluster NGC 6342, in the direction of the constellation Sagittarius (NGC stands for *New General Catalogue*, an index of about 8,000 objects far away in our galaxy). The peculiar thing about this pulsar is that contrary to most other binary pulsar systems, its rotational motion is not accelerated by the absorption of matter from its neighboring star.

Pulsars often have a very stable rotational period and would be very reliable timekeepers: some pulsars are just as precise as atomic clocks. Perhaps in the future a switch will be made — the cover of this book is a nod to the possibility of one day having a cosmic clock as our standard definition of time.

1.066,371 seconds
Rotational period of pulsar B2303+46

Little is known about this pulsar. Something is peculiar about it, though, because it appears to be accompanied by a white dwarf (see Chapter 2), which must have come into existence before the pulsar itself. Model calculations suggest that this binary system has had a complex history.

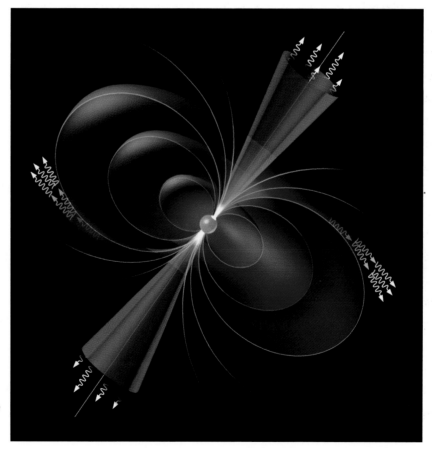

Epilogue

The shortest time span imaginable is practically incomprehensible to the human mind. This is the time a light ray needs to cross the 'Planck length' of 10^{-33} centimeters, which takes less than 10^{-43} seconds, also known as the 'Planck time'. On the other side of the spectrum is the time a light signal requires to cross the universe: approximately 50 billion years. This means that light would need 50 billion years to travel a distance equal to the size of the presently observable universe (but note that the universe started out very small and continues to expand). This is as far as we can travel in terms of distance, although timescales march even further on: there are small particles that take much longer to decay than 50 billion years, and the future of the universe can be extrapolated beyond 50 billion years as well. This is why the scales of time reach much further than the scales of distance. Our story ends only when we hit the 'dark eternities', regarding which there are so many uncertainties that it seemed wise to stop writing here.

Judging from our findings, we believe that the variety of time is much richer than distance. Gaping holes exist between various distance scales, where very few extraordinary phenomena occur. In contrast, we have seen that something remarkable happens at almost every timescale. Some phenomena occur only at one particular time period, others reappear everywhere. Electromagnetic vibrations can vary enormously in frequency. All these forms of radiation propagate with the exact same speed: that of light. We have also seen that there are many types of atomic nuclei and elementary particles that decay after a certain period of time, transforming into radiation or into other particles. This decay follows a set of almost universal laws, but the time each of them needs before this process is completed varies enormously — from the shortest measurable time spans to those much longer than the age of the universe.

We always made the assumption that the existence of so many differing timescales in the universe was natural, but is it really? Why are there so many timescales? To what do we owe this overwhelming complexity of the world in which we live? Not only do these questions provide fruitful ground for philosophers, there is also a practical angle, one that poses problems for physicists. This is because it touches upon the problem of 'unification'. How? Let us explain.

The descriptions we provide of natural laws governing our world are elaborate, but far from perfect. It is not difficult to pose questions to which we have no answers. It requires a whole lot more knowledge and expertise to ask questions that can indeed be answered through new experiments and sharp analyses. Above all, posing questions actually helps in building our understanding of things. The more we learn from our studies of nature, the more we identify connections between entirely separate phenomena. This, we call unification. And the unification continues. This is why we believe it must be possible to lead all natural phenomena back to a single all-encompassing law of nature, the 'Theory of Everything'.

Such a theory does not exist — yet. But it appears we are getting closer and closer. However, with our endeavors to formulate all-embracing natural laws, we encounter a remarkable problem, which is the issue we mentioned above: the extraordinary complexity of the universe. There are gigantic variances in scales of time and distance. Extremely strong and exceptionally weak forces coexist. What are the origins of these complexities? If we aspire to formulate an overarching theory, these intricacies must form a part of it, and this is exactly the reason why we have not yet been able to establish such a comprehensive theory.

Without this gargantuan diversity, life on Earth would never have evolved to what it is today — a complex culture of millions of totally unique organisms, with perhaps our society of (more or less) intelligent human beings representing an important pinnacle — for now.

Amazement

Fortunately we do not need an all-encompassing theory to marvel about the fascinating diversity of our universe. In this book we focus on the concept of time, and all that accompanies it. While passing through all imaginable timescales, we encounter many branches of science. We may well be scolded for describing the phenomena through the eyes of physicists, but we have made an honest effort to include several other types of science, including chemistry, biology, medical sciences, history, geology and astronomy. But when we reach the extremely long time periods, we return, sure-footed, to the familiar ground of physics. The circle is complete.

After we completed this book, we came across many other interesting subjects and wondered whether they should be included as well. Yes, indeed. Time moves on. Perhaps a later version of this book will be fleshier.

Time has a characteristic that cannot be found in space: time has *direction*. It is called 'the arrow of time'. By this we mean the evident fact that if you follow time into the future other laws appear to govern it than when you look back over the same period. The past is known to us, the future is not. How is this possible? Opinions vary, but as a matter of fact our world could never have existed in its current form if there had not been an arrow of time. Remarkably, though, at elementary, microscopic level the laws of nature appear to be symmetrical: if you show a movie of a falling rock or an elementary particle *in reverse*, the image is almost identical to the original. This is an important fact of the laws of nature, which, as you can imagine, generates a lot of discussion.

Will 'time travel' ever be possible? This is often a recurring theme in science fiction. Wouldn't it be beautiful if we could move to the future or back to the past? What bizarre situations would this ability create? Would it enable us to change undesirable aspects of the past to our every whim? It must be said that all laws of nature identified to date follow a strictly logical construction. This would be impossible to marry with the extremely *illogical* consequences of unbridled time travel, should it ever become possible. Most physicists do not believe time travel will ever be a realistic pastime. Quite the contrary: it has proven to be very useful to formulate existing laws of nature in a way that disallows sending signals back in time. Only then do our natural laws describe situations accurately, and this is why most physicists relegate time travel to the realm of the impossible — unless of course, this time travel occurs on paper or in our minds. This is exactly what we have done in this book. We are constantly amazed — hopefully you are as well.

Gerard 't Hooft and Stefan Vandoren

Glossary

Concepts in italics are mentioned separately in the list.

Alpha radiation

Radiation that consists of alpha particles, which are helium nuclei with 2 protons and 2 neutrons. During alpha decay of an *isotope* into another, alpha particles are emitted. It is one of the three types of radiation caused by *radioactive decay*. The other two are *beta radiation* and *gamma radiation*. Alpha radiation is the least harmful; even a piece of paper can stop this type of radiation.

Atomic clock

A device that can be used to measure time extremely precisely. The atomic clock is based on the vibrations of atoms with particularly precise *frequencies*. Atomic clocks are accurate up to less than one second per 30 million years. Today, most atomic clocks are based on the cesium atom (but other atoms could be used as well) and the time period of the second is defined using the vibration frequency of this element.

Beta decay

Radiation of radioactive decay that consists of beta particles. These are either negatively charged electrons (β^-) or positively charged positrons (β^+).

Electromagnetic waves

A wave is a signal that propagates with a particular velocity and *frequency*. Usually a wave is caused by vibrations or by disturbances in a medium around equilibrium, such as water particles going up and down in oceanic waves. Electromagnetic waves consist of vibrations of electric and magnetic fields. They do not require a medium and are able to propagate in vacuum. Light waves are a particular type of electromagnetic waves, with *frequencies* of between 10 (infrared) and 1,000 (ultraviolet) terahertz (1 terahertz = 10^{12} hertz).

Frequency

The number of times that a regularly repeated movement takes place within a certain period of time. The frequency of a system is 1 divided by the *period*. The frequency of electromagnetic radiation, such as light waves, is expressed in hertz (Hz), which indicates the number of vibrations per second. Visible light has frequencies varying between 430 (red) and 750 (violet) terahertz (1 terahertz = 10^{12} hertz).

Gamma radiation

One of the three types of radiation that can be caused by *radioactive decay*. Gamma radiation consists of high-energy gamma particles or photons (light particles). The energy of gamma radiation is higher than that of ultraviolet light, and can therefore be extremely harmful.

Half-life

The time it takes for a number of radioactive particles of a given kind to decrease by half through *radioactive decay*. After the half-life period only half of the initial number of radioactive particles remains.

Isotope

Isotopes are atoms that take the same position in the periodic table of elements — the table of Mendeleev — as the original elements to which they are related. They have the same chemical characteristics, but the structure of the atomic nucleus differs in number of neutrons. The number of protons (the atomic number) remains the same, and is the same as the number of electrons.

Light-second

This is the distance that light travels in vacuum in one second: 299,792,458 meters. Light-seconds are distances, and not timescales. A light-year is the distance light travels in vacuum in one year, which is 9,460,730,472,580,800 meters, or almost 10 trillion kilometres. A light-year consists of 31,556,736 light-seconds; in other words, the same number of seconds in a year.

Orbital period

The term orbital period is mostly used in astronomy. It is the time a satellite needs to complete a turn around the celestial body it belongs to. For instance, the Earth orbits the Sun in one year, the Moon orbits the Earth in about 28 days, and the Sun turns around the centre of our Milky Way in 220 million years.

Period

The period, or vibration time, of a periodic system is the time it takes for the system to return to its original state. The period of a pendulum is the duration of a completed oscillation. The period of the Earth's rotation is the time the Earth needs to turn around its own axis once, namely one day. The period of an *electromagnetic wave* is the duration of a complete vibration of the wave.

Quantum mechanics

A physics theory that describes the behavior and characteristics of atoms and electrons at microscopic (atomic and subatomic) scales. Traditional classical mechanics, based on Newton's laws, is no longer applicable at this scale. According to classical mechanics, for example, negatively charged electrons in atoms would be attracted by positively charged protons in the nucleus, causing the electron to end up in the nucleus. Quantum mechanics explains how it is possible that the electron remains at a fixed, minimal distance from the atomic nucleus, with energies that assume only certain discrete or 'quantized' values.

Radio waves

Radio waves form part of the electromagnetic spectrum. They are *electromagnetic waves* with frequencies between a kilohertz (10^3 Hz) and a gigahertz (10^9 Hz, 1 billion hertz). Radio waves are used in communication technology, to transmit information between antennae of senders and receivers, for instance radio and television.

Radioactive decay

A phenomenon caused by fundamental nuclear forces, the *strong and weak nuclear force*, and/or by the electromagnetic force. These forces are able to change the composition of an atomic nucleus, for example, by causing a neutron to decay into a proton via the *weak force*, or by causing a helium nucleus to split off from a heavier radioactive atom via the *strong nuclear force*. These processes release radioactive radiation, of which there are three types: *alpha*, *beta*, and *gamma radiation*.

Strong force

The strong force, also called the strong interaction, is — together with the gravitational, the weak, and the electromagnetic force — one of the four fundamental forces that act between elementary particles. It is responsible for holding the quarks together in the proton, and also for holding the protons and neutrons together inside the atomic nucleus. The strong force at low-energy experiments is much stronger than the other forces: it is about 100 times stronger than the electromagnetic force, a million times stronger than the *weak force*, and relative to the gravitational force it is 10^{38} times stronger. At higher energies, these differences become smaller. Despite its strength, the range at which the strong force acts is very small. It acts only at distances of the order of 10^{-15} meters, about the size of a proton.

Weak force

The weak and the *strong force* both act within the nucleus of an atom, but as opposed to the strong nuclear interactions, the weak force is much weaker and does not bind quarks or protons together. Instead, the weak force can transform a neutron into a proton, with the emission of an electron and and anti-neutrino. This process is also called *beta decay* and is responsible for radiactive processes. The weak interaction is the only force that breaks parity symmetry, a symmetry that would interchange left- and right-handed particles (the handedness of a particle refers to how it spins while it moves).

Literature and Websites

This book consists of hundreds of short stories, with a high degree of variation and a plethora of scientific themes. We have chosen not to provide source information for each section, because that list would be much too long. Moreover, in some instances it is difficult to find the original source, while it is not that relevant for the reader. Instead, we have listed a few of our favorite websites and books for further reading below (grouped in categories according to topic). In regards to the websites, it should be noted that exact addresses and contents change often, over which we have no control, but hopefully they point the reader in the right direction. Usually, multiple websites were used as sources, but those listed below would be a good starting point.

Powers of Ten

The original works about 'powers of ten' based on length scales can be traced back to:

1) Kees Boeke, *Cosmic View: The Universe in 40 Jumps*, John Day Co., Inc. 1957, ISBN: 0381980162. Translated into Dutch as *Wij in het heelal, een heelal in ons*, J.M. Meulenhoff — Amsterdam / J. Muusses — Purmerend, 1959.

2) The films of Charles and Ray Eames, *Powers of 10* (Vol. 1), 1968, Studio: Image Entertainment, DVD release date: August 15, 2000.

3) The movie is reprocessed in book form, in:

a) Philip Morrison, Phylis Morrison, and the Studio of Charles and Ray Eames, *Powers of Ten: A Book about the Relative Size of Things in the Universe and the Effect of Adding Another Zero*, W. H. Freeman and Company/Scientific American Books, 1982, ISBN-10: 0716714094, ISBN-13: 978-0716714095.

b) Original Dutch version: *Machten van Tien*, deel 1 van de Wetenschappelijke Bibliotheek, Natuur & Techniek, 1985, ISBN 9070157489.

4) The following Wikipedia website provides further information about the names and usage of large numbers: http://en.wikipedia.org/wiki/Names_of_large_numbers.

Timescales

There is a website with a movie on timescales. A manual can also be downloaded from that site:

Exploring time: http://exploringtime.org/?page=segments.

Wikipedia provides information about the orders of magnitude in time:

http://en.wikipedia.org/wiki/Orders_of_magnitude_%28time%29. It links to other pages about biological and geological timescales.

About extremely long timescales: Freeman Dyson, Time without end: physics and biology in an open universe, *Reviews of Modern Physics*, Vol. 51, No. 3, page 447, July 1979.

For a journey through time in steps of ten:

Nigel Calder, *Timescale*, Viking Press, New York, 1983, ISBN-10: 0670715719, ISBN-13: 978-0670715718

Isotopes

One of the book's recurring themes is that of the decay times of isotopes, set in pink borders. Lots of books exist that provide information on half-lives. We mention the following:

1) Richard B. Firestone (Author), S.Y. Frank Chu (Author), Coral M. Baglin (Editor), *Table of Isotopes: 1999 Update with CD-ROM*, Wiley VCH; 8th edition (9 August 1999), ISBN-10: 0471356336, ISBN-13: 978-0471356332.

2) Web links to the periodic table of elements, including information about isotopes can be found here:

http://www.webelements.com/

http://environmentalchemistry.com/yogi/periodic/

Astronomy

Data on orbital times and rotation periods of planets, comets, pulsars and galaxies are provided in many reference works. In our book this theme is indicated by the blue-colored boxes. Sources we have used include the websites of NASA and ESA:

NASA: http://www.nasa.gov/

ESA: http://www.esa.int/esaCP/index.html

These websites contain beautiful images and figures.

For more information about pulsars:

1) Duncan R. Lorimer, Binary and Millisecond Pulsars at the New Millennium,
   Living Reviews in Relativity 4 (2001) 5. URL: http://www.livingreviews.org/lrr-2001-5.

2) A list of binary and millisecond pulsars can be found in the appendix of this reference.

Elementary Particles

Information regarding elementary particles and their half-lives can be found in 'The Particle Data Book':

K. Nakamura *et al.* (Particle Data Group Collaboration), Review of Particle Physics, *J. Phys. G* 37, 075021 (2010). Web link: http://hepdata.cedar.ac.uk/lbl/pdg.html

Popular science books about elementary particles include the following:

1) Gerard 't Hooft, *In Search of the Ultimate Building Blocks*, Cambridge University Press, 1996,
   ISBN-10: 0521578833, ISBN-13: 978-0521578837.

   Dutch publication: *De bouwstenen van de schepping: een zoektocht naar het allerkleinste*, Prometheus, 1992, ISBN-10: 905333081X, ISBN-13: 978-9053330814.

2) Martinus Veltman, *Facts and Mysteries in Elementary Particle Physics*, World Scientific, 2003,
   ISBN-10: 981238149, ISBN-13: 978-9812381491.

   Dutch publication: *Feiten en mysteries in de deeltjesfysica*, Veen Magazines, 2004, Deel 78 van de Wetenschappelijke Bibliotheek, ISBN 9076988447.

General information about particle physics and the Large Hadron Collider at CERN, Geneva:

http://www.cern.ch/

Cosmology

A lot has been written about the history and evolution of the universe. This formed the green theme in our book. A few books we recommend for the general public include:

1) Stephen Hawking, *A Brief History of Time*, Bantam Books, 1988, ISBN 0-553-38016-8.

2) Steven Weinberg, *The First Three Minutes*, Basic Books, second edition, 1993, ISBN-10: 0465024378,
   ISBN-13: 978-0465024377.

3) Brian Greene, *The Fabric of the Cosmos*, Vintage, 2005, ISBN-10: 0375727205,
   ISBN-13: 978-0375727207.

Other Resources

As we noted in the Introduction, it was impossible to cover all topics comprehensively, but we did not want to keep the following interesting, remarkable and useful resources from our readers:

1) A general source for information, often a starting point of our research, is Wikipedia:
   http://www.wikipedia.org/

2) The bit about the history of the months July and August is based on:
   http://www.infoplease.com/spot/history-of-august.html#axzz0xPkblWSL

3) For more information about the mission of spacecraft New Horizons (Chapters 5 and 12), go to website:
   http://pluto.jhuapl.edu/

4) The source for traffic jams (Chapter 15) was: http://mobility.tamu.edu/ums/

5) Collision times and free path lengths (Chapter 38) can be found in: *Tables of Physical and Chemical Constants*, Kaye and Laby Online, section 2.2.4. http://www.kayelaby.npl.co.uk/general_physics/2_2/2_2_4.html

6) More information about thorium as energy source (Chapter 41) can be found at: http://www.itheo.org

7) The table of decay times and super heavy elements (Chapter 45) is based on: Yu. Ts. Oganessian *et al.*, Synthesis of a New Element with Atomic Number Z=117, *Phys. Rev. Lett.* 104, 142502 (2010).

8) An interesting piece about timescales and biology, from the millisecond to the circadian rhythm can be read here: Dean V. Buonomano, The Biology of Time across Different Scales, *Nature Chemical Biology* 3, 594–597 (2007).

Illustrations Credit

p. 39 University of Illinois. NCSA. http://archive.ncsa.illinois.edu/Cyberia/NumRel/EinsteinTest.html

p. 40 Wikipedia

p. 42 (top) Wikipedia

p. 42 (bottom) NASA/Wikipedia

p. 43 Marrhias Kabel/Wikipedia

p. 44 (left) http://heasarc.gsfc.nasa.gov/docs/cosmic/nearest_star_info.html

p. 44 (right) http://www.eso.org/public/outreach/eduoff/cas/cas2002/cas-projects/sweden_eridani_1/

p. 45 (left) Tim Malabuyo/Wikipedia

p. 45 (middle) Wikipedia

p. 45 (right) rebelpilot/Wikipedia

p. 45 (top) geni/Wikipedia

p. 46 NASA/Wikipedia

p. 47 (left) Wikipedia

p. 47 (top right) Wikipedia

p. 48 J. Linder/ESO/Wikipedia

p. 49 (left) The University of Manchester. http://www.jb.man.ac.uk/astronomy/nightsky/AList/Gemini.html

p. 49 (top right) NASA/Juan Carlos Casado (TWAN)

p. 50 Andrew Dunn/Wikipedia

p. 51 EightTNOs.png: Lexicon/Wikipedia

p. 52 (top) Matthew Field/Wikipedia

p. 52 (bottom) Gunnar Ries/Wikimedia Commons

p. 54 Robert A. Rohde/Wikipedia

p. 56 NASA/Wikipedia

p. 57 (top) Peter Brubacher. http://www.sacred-destinations.com/egypt/saqqara-step-pyramid-of-djoser

p. 57 (bottom) Berthold Werner/Wikipedia

p. 58 (top left) http://www.mnn.com/sites/default/files/styles/featured_blog/public/METHUSELAH.jpg

p. 58 (bottom right) Jon Sullivan/Wikipedia

p. 61 Wikimedia Commons

p. 63 Gerard 't Hooft

p. 64 Jose Luis Martinez Alvarez/Wikipedia

p. 65 Shutterstock

p. 66 ESA and the Planck Collaboration

p. 67 Shutterstock

p. 68 (left) Andrew from Cleveland, Ohio, USA/Wikipedia

p. 68 Wikipedia

p. 69 (bottom) Shutterstock

p. 71 (top) NASA/Wikipedia

p. 71 (bottom) John Moore. http://www.farnham-as.co.uk/2009/09/triangulum-galaxy-m33-imaged-by-john-moore/

p. 73 (top) Wikipedia. Dr. Ron Blakey Northern Arizona University. http/jan.ucc/nau.edu-rcb7/

p. 73 (bottom) Wikipedia

p. 74 (bottom left) Wikipedia/Heinrich Harder (1858–1935)

p. 75 Wikipedia/Chris Mihos (Case Western Reserve University)/ESO

p. 76 (right) Wikipedia/Berlin, Museum für Naturkunde, Axel Mauruszat

p. 76 (left) Wikipedia

p. 77   NASA/JPL-Caltech/R. Hurt (SSC/Caltech)

p. 78   Wikipedia

p. 79   Plymouth State University. http://
oz.plymouth.edu/~sci_ed/Turski/Courses/
Earth_Science/Intro.html

p. 81   NASA/Wikipedia

p. 82   Sea and Sky. http://www.seasky.org/

p. 84   NASA/Wikipedia

p. 86   Wikipedia

p. 88   http://www.physics.howard.edu/students/
Beth/bh_stellar.html

p. 89   Periodictableru/Wikipedia

p. 91   http://de.academic.ru/pictures/dewiki/84/
Tellurium_crystal.jpg

p. 92   Diego Grez/Wikipedia

p. 93   Kamioka Observatory, ICRR, The
University of Tokyo. http://www.
kennislink.nl/publicaties/ongrijpbare-
deeltjes

p. 95   http://newscenter.lbl.gov/wp-content/
uploads/abell-2218-mass-bends-light.jpg

p. 96   Alain r/Wikipedia

p. 97   http://astronomicando.blogspot.
sg/2008_02_17_archive.html

p. 98   NASA

p. 102   Wikipedia

p. 105   Gerard 't Hooft

p. 110   (top left) CERN. http://cds.cern.ch/
record/841555/files/lhc-pho-1998-349.jpg

p. 110   (top right) http://sbhep-nt.physics.sunysb.
edu/HEP/AcceleratorGroup/index.html

p. 111   CERN. http://public.web.cern.ch/Public/
features-archive/features/CMS%20
collision.jpeg

p. 112   http://sbhep-nt.physics.sunysb.edu/HEP/
AcceleratorGroup/index.html

p. 113   http://sbhep-nt.physics.sunysb.edu/HEP/
AcceleratorGroup/index.html

p. 114   Gerard 't Hooft

p. 115   (top left). Gerard 't Hooft

p. 123   (top) NASA

p. 124   Qwerter/Wikipedia

p. 125   (top left) http://www.odec.ca/
projects/2004/sitt4b0/public_html/images/
tungstencrystal.jpg

p. 125   (top right) Indian Institute of Technology
Kanpur. http://home.iitk.ac.in/~sreerup/
bso203/debyscherrer.jpg

p. 125   (bottom) Gerard 't Hooft

p. 127   Deutsches Röntgen Museum. http://www.
roentgenmuseum.de

p. 128   (right) Magee-Women's Hospital of
UPMC. http://well.blogs.nytimes.
com/2008/04/10/ mammograms-new-
and-old/

p. 130   Wikimedia Commons

p. 132   Wikimedia Commons

p. 133   (top left) http://www.
creaseymahannaturepreserve.org/
flirtatious-flowers/

p. 133   (right) Max-Planck-Institut für Physik
komplexer Systeme. http://www.mpipks-
dresden.mpg.de/~atto07/

p. 135   http://www.astronomy.ohio-state.
edu/~pogge/TeachRes/Ast161/Atoms/
SunSpectrum.jpg

p. 136   Shutterstock

p. 137   https://dlnmh9ip6v2uc.cloudfront.
net/assets/8/c/5/2/1/511917bbce395f
ef32000000.jpg

p. 139 (top) InfoEscola. http://www.infoescola. com/quimica/femtoquimica/

p. 139 (bottom) JabberWok and Time3000/ Wikipedia

p. 145 Shutterstock

p. 148 (top right) Krzysztof Szymański/Wikipedia

p. 149 (top) Buienradar.nl

p. 150 Wikipedia

p. 152 (top left) http://www.maclife.com/tags/ cell_phones

p. 153 (top right) Shutterstock

p. 155 CERN. The European Organization for Nuclear Research

p. 156 http://tymkrs.tumblr.com/ post/4838083914/21-ionosphere-ham-lesson-o-de-day

p. 158 Michael Maggs & Richard Bartz/Wikipedia

p. 159 (top left) Andrew Davidhazy/Rochester Institute of Technology

p. 159 (middle) http://periodictable.com/ Isotopes/089.218/index2.p.full.html

p. 160 (top left) http://mrgraves9hmrsci. wikispaces.com/Mitch+C+-+Actinium

p. 162 Shutterstock

p. 163 (right) Wikipedia

p. 165 http://www.engadget.com/2007/05/08/ europes-galileo-satellite-navigation-system-at-a-dead-end/

p. 166 Ideru/Wikipedia

p. 167 (top) Wikipedia. Musée International d'Horlogerie, La Chaux-De-Fonds, Switzerland

p. 167 (bottom) Goudsmederij Mazlemian. http://mazlemianbros.nl/N_xp26_ Memo_7.gif

p. 168 Deutsch Wikipedia. de.academic.ru http:// de.academic.ru/dic.nsf/dewiki/1440952

p. 170 (right) Gerard 't Hooft

p. 171 Wikimedia Commons

p. 172 http://www.mpe.mpg.de/1048427/ CompactObjects

p. 173 Borb/Wikipedia

p. 175 (top) Alan Vernon/Wikipedia

p. 175 (bottom) Shutterstock

p. 176 Shadowfax/Wikipedia

p. 177 (top) Shutterstock

p. 177 (right) The Cornelllab of Ornitology. http://www.birds.cornell.edu/brp/ elephant/cyclotis/language/dictionary.html

p. 177 (bottom) Shutterstock

p. 179 (top right) http://www.aviationnews. eu/2008/03/10/lockheed-martin-submits-proposal-to-provide-goes-r-spacecraft/

p. 180 (left) Norman Bruderhofer/Wikipedia

p. 180 (middle) http://www.cottoneauctions.com/ displayItem.php?displayItem_id=6150

p. 180 (right) Atreyu/Wikipedia

p. 180 (top right) Paulnasca/Wikipedia

p. 181 Shutterstock

p. 183 http://avengers-in-time.blogspot. sg/2012/07/1967-science-technology-pulsar.html

Index